A QUESTION of EVIDENCE

The Casebook of Great Forensic
Controversies, from Napoleon to O. J.

Colin Evans

John Wiley & Sons, Inc.

Library of Congress Cataloging-in-Publication Data:

Evans, Colin, date.
    A question of evidence : the casebook of great forensic controversies,
  from Napoleon to O. J.
       p. cm.
    Includes bibliographical references and index.
    ISBN 0-471-44014-0 (acid-free)
    1. Criminal investigation. 2. Criminal investigation—Case studies.
  3. Forensic sciences. 4. Evidence, Criminal.   I. Title.
  HV8073 .E926   2003
  363.25—dc21   2002014326

Contents

Acknowledgments

Among the many who gave so generously of their time and help while this book was being written, the following deserve special mention: Tony Cassidy of the Thames River Police; Susan L. Dow; Ronald Mirvis; Kenneth Rudolf; Matt Scallan; Jennifer Schizas; Sam Shaw; Pat Sloan; Pat Small; David Tyler; Ann Walding-Phillips; and Piers Windsor. I am particularly indebted to Linzey Coles at King's College, London, for her research assistance.

The staff at the British Library and the Newspaper Library at Colindale were unfailingly helpful, as were the countless reference librarians who answered my questions and steered me in the right direction.

Once again, David Anderson put his personal library at my constant disposal, whilst Greg Manning proved to be a mine of information on the Calvi case.

As always, thanks to my agent, Ed Knappman, and everyone at NEPA.

Finally, a special note of appreciation to the people at Wiley: Kimberly Monroe-Hill and William Drennan for the fine way they handled the copyediting; and Jeff Golick, who first suggested this book, and whose input and comments were such an integral part of its creation.

Introduction

The evolution of forensic science has been a long, complex, and fascinating journey. For the most part it is a story of triumph, a succession of victories—some large, others barely noticeable—in the never-ending battle to close the loopholes through which criminals slip. This progress has been exponential. Although the first tenuous steps in scientific crime solving came as early as the eighteenth century, the giant strides didn't really happen until after World War II. Not only did the atomic age bring about a quantum shift in technological development—there is nothing quite like the threat of mutually assured destruction to concentrate the scientific mind—but also, more significantly, crime had replaced war as the number one social evil in the Western world. There were many reasons for this: suddenly, society was swamped by an influx of millions of homecoming ex-servicemen, for whom violence had become a way of life; increased affluence meant that householders now had more possessions worth stealing than ever before; while increased mobility, in the form of affordable vehicles, gave the enterprising lawbreaker a vastly enhanced workplace. Now, as crime hit closer to home, all those two-bit hoodlums whose overhyped exploits had enlivened Depression-era newspaper headlines no longer seemed quite so appealing. With crime levels soaring through the roof, it became blindingly obvious that the old standbys of crime detection—shoe leather, informants, and methodical elimination (though these still form the bedrock of most investigations)—were not enough to stem the onslaught. New weapons were needed.

1

Enter the crime lab. Suddenly the electron microscope, the spectrograph, gas chromatography, DNA typing, and a hundred and one other subbranches of the forensic detection tree became indispensable allies of the investigating officer.

Only an imbecile would gainsay the enormous benefits that forensic science has brought to the business of crime-solving. Without it, incalculable numbers of thugs nowadays behind bars would otherwise be roaming the streets, and for that we should all be truly grateful. It was to chart this progress that I wrote *The Casebook of Forensic Detection,* and judging from the response of readers, it was an approach that met with considerable approval. However, among the many kindly letters I received were one or two expressing concerns about the downside of all this progress. Weren't we in danger of placing too much faith in the word of a white-coated scientist just because he wears a white coat? Surely there must have been an occasional blunder? After all, nobody bats a thousand—do they? This set me thinking, and the book you now have before you is the culmination of those deliberations.

The cases found within these pages are the great imponderables, the big beasts of the forensic science jungle, all guaranteed to generate plenty of heated discussion. They range from the Middle Ages to current times. Most are well known, a few less so, but all are crammed with scientific controversy, whether it be through botched experiments, faulty data, blatant evidence tampering, hubris, or just plain pigheadedness. This last trait is surprisingly common, and makes nonsense of the concept that the duty of science is to eliminate emotion from the equation. People may be scientists by training, but they are humans by instinct, and subject to all the frailties that that involves. For too many expert witnesses, this takes the form of intransigence at all costs. Once they have set out their evidentiary stall, they will not budge an inch, no matter how compelling evidence to the contrary may be. Not for them the dictum of the great economist John Maynard Keynes, who responded to taunts of inconsistency by famously firing back, "When the facts change, I change my mind. What do you do, sir?"*

In many respects, forensic science has become a victim of its own success. Nowadays it has a virtual stranglehold on the upper echelons of the legal system. In almost every serious trial, both sides want to get in as much favorable science as possible, because they know just how popular it is with juries. They know that the men and women who will decide this case have,

---

in all probability, read articles in magazines and newspapers, or watched at least one of the numerous TV programs devoted to crime-solving, and they will have marveled at the undeniably spectacular results achieved by science. A drop of blood, a speck of dust almost invisible to the naked eye, discerning the life cycle of some insect as an aid to establishing time of death—science can analyze all of these and more. Nothing seems to be beyond the reach of the modern crime laboratory.

As reassuring as these developments may be, with such progress comes danger—the tendency of juries to accept, without question, what an expert witness tells them. Science is supposed to solve crimes, not aid them; and yet, as we shall see, even the greatest experts are far from infallible. On one occasion, misguided testimony from one medico-legal giant, regarded as the supreme authority of his age, not only allowed a brutal killer to walk free but also led directly to the murder of two more women.

Mistakes are one thing, crookedness is another, and a disturbing theme that does keep recurring is expert witness malleability. "If the law has made you a witness, remain a man of science,"* wrote Dr. P. C. H. Brougardel, a late-nineteenth-century forensics expert. Wise words; but as the reader will discover, there is plenty of evidence in this book to suggest that in the fiercely competitive and lucrative world of the expert witness, testimony frequently owes more to whoever is cutting the check than it does to impartial analysis.

Corruption is nothing new. In the latter half of the nineteenth century, as science was gaining a foothold in the courtrooms, it became clear that juries were often overawed by the mere presence of some "doctor" with a microscope and a few photographs, and there were plenty of hucksters prepared to cash in on this gullibility. Nowadays some of the forensics advances, particularly in the area of computer CAD technology, with its putative attempts to re-create certain crime scenes, and in the legal minefield that is psychological profiling, are just as troubling. For no matter how fancy the trappings, both disciplines, when stripped to their essentials, are nothing more than speculation masquerading as science, and should be regarded as such.

Selecting which cases to discuss has not been easy. I have deliberately steered away from the dozens of tragic miscarriages of justice that DNA typing has unearthed in recent years (six inmates freed from American death rows in 2000, alone) and concentrated, instead, on cases where a strong element of doubt still exists. There are two reasons for this. First,

---

*Milton Helpern and Bernard Knight, *Autopsy* (London: Harrap, 1979), p. 65.

they're more interesting; and second, this is a book about controversies. You won't find any sterile certainties here. What you will find are profound misgivings and disputes, bitter feuds, and the ways in which strongly held passions—for both good and evil—can still occasionally vanquish the best that science has to offer. Because many of the cases are ongoing, I fully expect that some readers will form views widely different to my own, but that's the nature and the appeal of controversy, and why this book was written.

Chapter 1

The Turin Shroud (1355)

Genuine Relic or Medieval Fake?

Hard-core criminality is nothing new. Catch the evening news, with its endless litany of drive-by shootings, armed holdups, and other scenes of urban mayhem, and it's all too easy to run away with the notion that we are living in the most lawless era in history. Nothing could be farther from the truth; we're just better informed. For sheer, unadulterated havoc and skulduggery, nothing can top the Middle Ages. Setting aside the blood-drenched wars and the ravaging plagues, it was a time when murder was commonplace; rape went unpunished; and incest thrived in remote rural areas. It was an age of quite extraordinary violence to person and property, and wholesale theft. From the highest in the land to the lowliest peasant, nobody escaped the ravages of crime—not even the established church.

In the Middle Ages the Roman Catholic Church was a vast, sprawling corporation, the biggest business on earth; like most multinationals, it was riven by internecine feuding. Not all of the disputes were theological; greed played its part. For while it's true that all financial roads may ultimately have led to Rome, across Europe hard-nosed priests were after their slice of the cake as well. Competition for the franc, florin, ducat, guilder, and groat was razor-sharp, with individual parishes fighting hard to attract the biggest congregation. Everyone was looking for an edge, and the biggest edge of all was undoubtedly a religious relic. Any house of worship that could boast within its precincts some artifact with a biblical connection or, better still, a scrap of human matter, was on the high road to hitting the pilgrim jackpot. Mere rumor of such a treasure was enough to bring the faithful flocking in droves. And with these pilgrims came money; oodles of the stuff. As a result, the most celebrated of these shrines developed into medieval theme parks, generating huge income streams, not just for the church but for the surrounding microeconomy as well. Needless to say, with so much money sloshing around, it was only a matter of time before someone decided to prime the relic pump. The result was a forger's bull market.

From the Mediterranean to the Baltic and across to Iberia, "relics" started sprouting up everywhere: crowns of thorns; assorted Holy Lances; even a foreskin or two from Jesus' circumcision; and enough slivers of the True Cross to rebuild Noah's Ark. In Loreto, Italy, they went one better, boasting an entire Nazarene house allegedly once inhabited by Jesus, Mary, and Joseph, all flown there by angels in 1294. The twelfth-century Three Kings Cathedral in Cologne, Germany, houses bones purportedly belonging to the Magi who visited the infant Jesus. Another German city, Trier, plays host to the Holy Coat, the seamless tunic worn by Jesus at his crucifixion; unfortunately, the parish church at Argenteuil, just north of Paris, claims the same distinction.

When it comes to the single-minded pursuit of a desirable relic, pride of place must surely go to St. Hugh of Lincoln, England. The story goes that when shown an arm of St. Mary Magdalene in Provence, he attempted to remove a chunk with his knife to take back to his cathedral. When that failed, he gnawed some off and brought the morsel home. Such macabre souvenir-hunting did little to detract from his own sanctity, and his magnificent tomb in Lincoln Cathedral was itself a popular site of pilgrimage until it was despoiled during the Reformation.

Throughout medieval Europe, trickery was rampant. Relics were stolen, fenced, copied perhaps dozens of times, then slipped back into a hungry market. The undoubted scam capital of the age was Venice. As the greatest sea power in the Mediterranean, Venice controlled the major east–west trade routes, and its craftsmen were legendary for their ability to replicate treasures plundered from Asia Minor, which were then trafficked all over Europe.

Some relics may have been genuine. Most, palpably, were not. Yet their marketability remained unaffected. Because people tend to believe what they want to believe, unhampered by facts or reason, the charlatans were never in danger of going out of business. This trend is still with us. In 1988 Monsignor John Ellis, a Catholic Church historian, recalled with distaste his visit to St. Anthony's Church in Padua, Italy, where the guide glibly pointed out a vial containing "the milk of the Blessed Virgin Mary." As Ellis remarked, "Some relics are fed by sheer curiosity, but some are by fanaticism. I don't say there are no real relics, but there's so much fraud you can't be sure."[1]

For six centuries and more, the undoubted king of relics has been the Turin Shroud. At a little over fourteen feet by three feet, it is also one of the largest. Superbly made from fine herringbone twill linen, it is, so millions believe, the actual cloth used to wrap the body of Jesus after it was taken down from the Cross and placed in a tomb. In making this claim, the Turin Shroud isn't unique—researchers have identified about forty other alleged "shrouds" throughout the world—but what makes this particular specimen so special is its detail. Alone amongst all the claimants, the Turin Shroud actually depicts, in faint markings on a sepia background, a discernible human image.

Since it was moved to Turin in the sixteenth century, the Shroud has been an object of awe and reverence for generations of devout Catholics. They came in the millions to pray and marvel. On some days the crush was so great that pilgrims died from suffocation. But it was the events of May 28, 1898, that escalated the Shroud from devotional icon into an object of universal curiosity.

A prominent Turin councilor, Seconda Pia, had been commissioned to make the first official photograph of the Shroud, which hung in the cathedral. Working at night to avoid the crowds, he wrestled with his cantankerous equipment and the poor lighting until he eventually succeeded in capturing two plates of the faint image, front and dorsal. Harboring no great expectations of a successful outcome, Pia returned to his darkroom to develop the plates. There, to his astonishment, he found that when he examined the plates, the negative revealed an infinitely more lifelike image of the figure on the cloth than was visible with the naked eye.

The detail was remarkable: a nude, bearded male, almost six feet tall, with long hair, lay with his eyes closed, hands crossed over the groin, and his right foot slightly raised. There appeared to be a large, open wound in the chest near the heart, what looked like injuries to the wrists and feet, and a fretwork of lacerations across the back, such as might have been caused by scourging. One didn't need to be a biblical scholar to realize that these injuries tallied unerringly with those recorded in Gospel accounts of the crucifixion of Jesus.

For Pia, the moment was an epiphany, and for the rest of his life he was convinced he had looked upon the face of Jesus, newly killed on the Cross. Reports of his discovery flew around the city, then the world, with

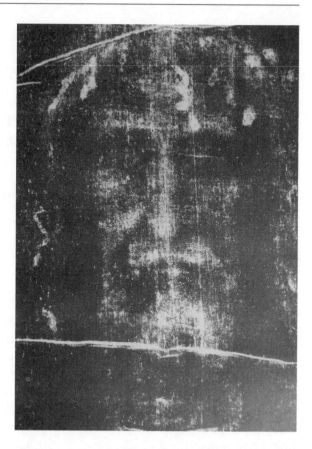

The Turin Shroud: Is this the
face of Jesus Christ?

the result that even greater crowds besieged the cathedral, clamoring to glimpse what most regarded as a miracle. Others, religious leaders and scientists mostly, fearing a hoax, subjected Pia to a withering interrogation, but the doughty councilor stood by his results.

So was this a genuine miracle, or was it, as many insist, the biggest and most enduring art fraud in history?

The first person to address this conundrum scientifically was Yves Delage, a distinguished French professor of anatomy and a declared agnostic. For eighteen months, he and his assistant, Dr. Paul Vignion, pored over both the plates and the actual cloth, studying every square millimeter. Delage revealed his findings in a lecture given on April 21, 1902, at the Académie des Sciences in Paris, saying that, in his opinion, the Shroud body image and wounds were so physiologically flawless that he found it impossible to believe they could be the work of an artist. Moreover, microscopic examination of the cloth revealed what he believed to be clear evidence of blood. The human impression, he said, had been caused by the urea of body sweat, and spices used to anoint a dead body. To breathless silence in the auditorium, Delage reached his dramatic crescendo, a declaration that he found no difficulty in believing that the body wrapped in the Shroud was that of Jesus. A huge burst of cheering greeted the announcement.

One would imagine that the church's reaction to such an unqualified endorsement would have been unalloyed joy, but it wasn't that way at all. When it comes to the Turin Shroud, the Vatican has always been jittery, preferring to keep a certain distance, perhaps fearful that here was a religious banana skin just waiting for an unwary foot.

Origins of the Shroud

To understand this ambivalence, we need to travel back to the mid-fourteenth century, and the first authenticated mention of the Shroud. In 1389 the then bishop of the French city of Troyes, Pierre d'Arcis, wrote angrily to Pope Clement VII in Avignon about a scandal he had unearthed in the tiny church of Lirey, which lay within his diocese. Much to his fury, the church's canons had "falsely and deceitfully, being consumed with the passion of avarice and not from any motive of devotion but only of gain, procured for their church a certain cloth cunningly painted, upon which by clever sleight of hand was depicted the twofold image of one man, that is to say the back and front, they falsely declaring and pretending that this was the actual Shroud in which our Savior Jesus Christ was enfolded in the tomb."[2]

In trashing the Shroud, d'Arcis stepped on a lot of toes. Lirey was in the midst of a relic bonanza, with pilgrims snapping up medallions as mementos of their visit. Fortunes were being coined and such was the wealth in circulation that it led to murmured complaints of financial jealousy against d'Arcis, grumpy because he wasn't getting his cut. This makes no sense whatsoever. As bishop of the diocese his power was virtually limitless. Had he wanted to sequester, if not all, then at least a sizable chunk of Lirey's Shroud-based income, doubtless he could have. Instead, he seems to have wanted no part of the relic.

He amplified his concerns in the memo. The cloth, he wrote, had first been shown in Lirey in about 1355, at the time of his predecessor, Bishop Henry de Poitiers. When the matter was brought to his attention, Poitiers made inquiries and "discovered the fraud and how the said cloth had been cunningly painted, the truth being attested by the artist who had painted it, to wit, that it was a work of human skill, and not miraculously wrought or bestowed."[3]

Since even the de Charny family—the Shroud's original owners and, coincidentally, the founders of the church at Lirey—referred to the item in documents, not as a burial cloth of Jesus, but merely as a "likeness or representation,"[4] ordinarily that would have been an end to the story. But in 1453 the cloth was willed to the Italian royal house of Savoy, and it was this journey to another country that really brought about a change in the Shroud's fortunes and iconic status.

Just eleven years later, the future Pope Sixtus IV felt confident enough to describe it as "colored with the blood of Christ,"[5] and it wasn't long before the Shroud was even granted its own feast day. A brush with calamity came in December 1532, when a fire threatened to destroy the Shroud as it lay in a silver reliquary, locked behind a grille. Only quick work by a local blacksmith, forcing the bars and snatching the box from the flames, saved the day. Though the Shroud suffered some fire and water damage, this was patched in 1534 by a team of nuns, who also strengthened the relic adding a backing cloth. In 1578 the Shroud ended its travels when it was transferred to a specially built chapel in Turin Cathedral, where it remains to this day.

Actual ownership of the Shroud only passed into the hands of the Vatican in 1983, and since then, on those rare occasions when it has been removed from its wooden casket and displayed for all to see, it has continued to exert a magical spell. During the last showing, in 1998, an estimated three million visitors filed past in just eight weeks, with few of those spectators prepared to question the Shroud's authenticity. They prefer to leave that argument to others.

For the most part the debate has been bilious and incredibly hostile. Normally phlegmatic, rational scientists have a tendency to work themselves into a monumental lather when it comes to the Turin Shroud, no matter which side of the fence they're on. All of which makes sifting reliable evidence from the invective no easy task.

The first new evidence since the efforts of Yves Delage came in the 1970s, with a sensational announcement from a Swiss criminologist, Dr. Max Frei-Sulzer, that pollen grains he had scraped from the Shroud's surface had come from no less than fifty-eight different Middle East plant species. Because pollen grains are extraordinarily durable—they can survive for thousands of years—this makes them an extremely useful forensic archaeology tool, and Frei-Sulzer's pronouncement appeared to deal a knockout blow to those jeering skeptics who maintained that the cloth had never traveled any farther east than the Franco-Italian border.

Unfortunately for Frei-Sulzer, his credibility as an analyst plunged like a lead parachute in May 1983 when the notorious Hitler Diaries that he had "authenticated" were exposed as a flagrant hoax.* Death in March of that same year had spared Frei-Sulzer the embarrassment of the diaries fiasco, but did nothing to stifle persistent whispers that the flashy criminologist-cum-botanist from Zurich had resorted to spiking the Shroud sample with pollen personally acquired from repeated trips to Turkey and the Holy Land, though this was never proved.

Even before this revelation, the skeptics had begun to fight back. Leading the charge was Dr. Walter C. McCrone, an outspoken microanalyst from Chicago, who, in 1979, attempted to chemically analyze the image. Using more than two dozen samples taken from the Shroud with sticky tape, McCrone subjected them to a full range of forensic tests. His findings were delivered with customary bluntness: "There is no blood on the Shroud."[6] What he had found, though, was paint. Through the use of polarized light microscopy, he had identified what he believed to be clear traces of the pigment vermilion, in addition to red ocher and tempera, all paints in common use during the Middle Ages. Armed with this knowledge, an artist friend of McCrone's, Walter Sandford, managed to produce a passable Shroud-like image, providing at least partial confirmation for McCrone's view that the Shroud was a "beautiful medieval painting."[7] As there had been no reference to the Shroud before 1355, McCrone concluded that the Shroud had probably been painted shortly beforehand, then stored "to give the paint a year to dry"[8] before being sold to the church in Lirey.

Predictably such an emphatic renunciation brought an avalanche of criticism crashing down on McCrone's head, with accusations of sloppy methodology, grandstanding for the media, and an anti-Christian agenda. Some of the criticism was colorful, some was libelous, and it should be said that McCrone gave as good as he got in the acrimonious exchanges, but the most telling outcome of all this verbal sparring was the way it spurred other scientists into action. In 1981 Dr. Alan Adler, a renowned chemist, fired back a string of equally impressive test results, which did appear to show the presence of blood on the Shroud. However, even if true, there is nothing to say that this blood is two thousand years old, and plenty to suggest that it isn't.

When the Shroud suffered fire damage in 1532, only the fact that it was housed in a silver reliquary saved it. Even so, it was still scorched.

---

*For a full account of this debacle see Colin Evans, *The Casebook of Forensic Detection* (New York: John Wiley & Sons, 1996).

Contemporary accounts record that the fire's intensity melted some of the metal, causing a drop of molten silver to land on the cloth (the mark is visible to this day). As the melting point of silver is 961°C—approximately the temperature used to cremate bodies—such extreme heat would drastically affect any blood proteins found on the cloth and compromise their analytical worth. Even if, as supporters—they call themselves Shroudies—claim, the cloth was folded and therefore slightly insulated from the worst of the heat, such a roasting must have taken its toll.

For this reason, what blood—if any—there is on the Shroud probably originated after the fire, most likely from one of the countless sick pilgrims who must have touched the Shroud over the centuries in hopes of a cure.

So if the scientists couldn't agree on what the Shroud revealed chemically, what would a forensic pathologist be able to make of the image itself? At the request of the British Society for the Turin Shroud, Professor James Cameron, a man with considerable experience of violent death, brought his talents to bear on the puzzle.

Considering the paucity of material he had to work with—just a few photographs—some of his conclusions were eye-poppingly expansive. "The image of the face is indicative of one who has suffered death by crucifixion,"[9] he wrote, without explaining in what way the face of a crucified person differs from that of someone who has suffered any other kind of death. Then, turning to the dorsal image, he discerned evidence of "deep bruising of the shoulder blades, indicating the angle at which the cross beam of the Cross might have been carried."[10] He also noted that "the scourge marks on the body would be consistent with a flagrum,"[11] a particularly nasty short-handled Roman whip that had pellets of bone or lead attached to its thongs.

Cameron went on: "The image on the Shroud indicates to me that its owner—whoever he may have been—died on the cross, and was in a state of rigor when placed in it."[12] Given the sweeping scope of his previous observations, it came as something of an anticlimax when he rather tamely concluded, "It is my belief that we shall only be able to prove . . . that the Turin Shroud might be the burial cloth of Jesus Christ, not that it actually is."[13]

The forensic battle heated up. Scraps of "evidence" were tossed into the fray by both sides, and just as quickly shot down by the opposition. Finally, in 1988, came a breakthrough. In a surprise move, the Vatican gave permission for minute samples from the Shroud to be subjected to radiocarbon dating. Here at last, or so it appeared, the mystery would be solved once and for all.

How Radiocarbon Dating Works

Radiocarbon dating is the most widely used method of age estimation in the field of archaeology, and works by means of measuring the amount of carbon 14 left in an object. The principle was pioneered by Willard F. Libby at the University of Chicago in the 1950s, working with items of known age. This groundbreaking research earned Libby the 1960 Nobel Prize in Chemistry.

Certain chemical elements have more than one type of atom, and different atoms of the same element are called isotopes. Carbon has three main isotopes: carbon 12, carbon 13, and carbon 14. Of these isotopes, carbon 12 is the most abundant, making up 99 percent of the carbon on Earth; next comes carbon 13 at 1 percent; and right down at the bottom of the scale is carbon 14, which makes up just 1 part per million. What makes carbon 14 so useful is that alone among these three isotopes, it is radioactive, and the gradual decay of this radioactivity is used to measure age.

Radioactive atoms decay into stable atoms by a simple mathematical process. Half of the available atoms will change in a given period of time, known as the half-life. For instance, if 1,000 atoms in 2000 had a half-life of 10 years, then in 2010 there would be 500 left. In 2020 there would be 250 left, in 2030 there would be 125 left, and so on.

Therefore, by counting the number of carbon 14 atoms in any object that contains carbon, it is possible to calculate either how old the object is, or how long ago it died. For this we need to know two things: the half-life of carbon 14; and how many carbon 14 atoms the object had before it died. The first part is straightforward: the half-life of carbon 14 is 5,730 years. However, knowing how many carbon 14 atoms something had before it died can only be estimated, but the assumption is that the level of carbon 14 in any living organism is constant—that is, when a particular fossil was alive, it had the same amount of carbon 14 as the same living organism today.

Dates derived from carbon samples can be carried back about 50,000 years. Beyond that, it is necessary to employ potassium or uranium isotopes, which have much longer half-lives. These are used to date very ancient geological events that have to be measured in millions or even billions of years.

This, then, was the technology. And on April 21, 1988, under the watchful gaze of Cardinal Ballestrero of Turin and a video camera, Italian microanalyst Giovanni Riggi cut a 1/2-inch by 3-inch strip of linen from the Shroud, well away from its central image and any charred or patched areas. He divided the strip into three postage stamp–size samples and distributed them to representatives of laboratories in Zurich, Oxford, and Tucson. Each then performed at least three radiocarbon measurements on its sample.

Because of the small sample size, the method of measurement used was accelerator mass spectrometry (AMS). Although at that time AMS was a less well-developed technique than that used for the majority of radiocarbon datings, the participating laboratories had already measured several thousand dates by the AMS method, and its accuracy, both then and subsequently, has been shown to be comparable to the best of the laboratories using conventional methods.

All three laboratories, working independently and with controls, came up with the same result: the linen cloth used to make the Shroud was manufactured between A.D. 1260 and 1390, a time frame that covers exactly the Shroud's first known appearance in Lirey.

The Oxford team made their announcement at a British Museum press conference, stating that the radiocarbon dating results "provide conclusive evidence that the linen of the Shroud of Turin is medieval."[14] According to nuclear physicist Harry Gove, the odds were "about one in a thousand trillion"[15] against the Shroud's having been woven in the time of Jesus. Edward Hall, another member of the Oxford team, decided to twist the knife. Anyone who continued to believe the Shroud was genuine, he jeered, must be a "flat-Earther."[16]

Shroudies were stunned. At a stroke, centuries of unquestioning belief appeared to have been turned to dust.

But zealots are hardy souls, and soon the hunt was on to find some way to discredit the findings. Hysterical accusations that the laboratories had colluded in an atheistic conspiracy to rig the results were treated with appropriate contempt and when off-the-wall claims of sample switching started circulating, Robert Hedges, of the Laboratory for Archaeology at Oxford, just shook his head in disbelief. "Having witnessed the sampling operation, I find this assertion incredible."[17]

Yet another noisy faction alleged that the samples had actually been cut from the backing cloth sewn on by nuns in 1534, and this accounted for the misleading results; though quite why the nuns used two-hundred-year-old cloth to patch the Shroud was not made clear, but by this stage some supporters were getting distinctly panicky.

Others, more rationally, zeroed in on the reliability of carbon dating, and cited instances where it had been shown to be inaccurate, though never by such a margin on such a recent artifact. For the carbon dating to be off by thirteen hundred years, something must have been radically wrong with the sample. Soon, though, relieved Shroudies thought they'd found the answer.

In 1996 a team of researchers at the University of Texas announced that an alleged sample of Shroud fibers they had studied was mired in bioplastic contamination. This occurs when living organisms—bacteria and fungi, typically—add carbon of a "fresher" nature to an existing sample. If not thoroughly cleaned off, this contamination could distort any findings toward a much later date, which is what happened here, they claimed. Instead of sampling the fiber's cellulose alone, the 1988 testers had sampled the contaminants as well.

Again Hedges was scathing, noting that "the degree of contamination required to shift a thirteenth-century date by thirteen hundred years is very large (such a shift would require the addition of about 50 percent more material of "modern" carbon), and this quantity, or indeed any amount above a few percent, can be totally ruled out."[18]

Further doubts about the Texas results surfaced from an unusual and authoritative quarter when Turin's Cardinal Giovanni Saldarini, custodian of the Shroud, publicly questioned the origins of the Texas sample. On Italian television in 1996 he was quoted as saying: "There is no certainty that the material belongs to the Shroud so that the Holy See and the custodian declare that they cannot recognize the results of the claimed experiments."[19]

But Dr. Leoncio Garza-Valdes, the microbiologist who headed the Texas team, was nothing if not persistent, and in 1998 he was back, this time claiming that red areas on the Shroud, far from being paint, as McCrone and others had argued, were actually ancient blood stains. Furthermore, he declared, the type AB blood was "common among Jewish people."[20] This was a baffling statement. Not only is AB not especially prevalent in Jewish people, but according to a noted serologist, Dr. Peter D'adamo, AB is a "new" blood, probably caused by the intermingling of type A Caucasians with type B Mongolians in the fourth to seventh centuries A.D. There is no evidence for its existence beyond approximately a thousand years ago. Undeterred, Garza-Valdes plugged on, giving hope to Shroudies everywhere, by further claiming to have discovered a few fragments of oak—a common tree in Jerusalem—on the cloth, heightening speculation that these might have come from the Cross.

One year later, the question of pollen found on the Shroud was raised once again, this time by two Israeli experts, Dr. Avinoam Danin, professor of botany at Hebrew University in Jerusalem, and Dr. Uri Baruch, an expert in pollen dating, at the Israel Antiquities Authority. Danin said the most common pollen was that of the plant *Gundelia tournefortii*. "This is a thorny plant described in the Bible as tumbleweed. Some Christians say it formed the crown of thorns. It grows only in the Near East. Therefore the Shroud could only have come from the Near East."[21] Other pollen found on the Shroud is that of the bean caper, which is found between Jerusalem and the Jordan Valley.

It might be possible to attach more weight to these findings had they accrued from fresh tests, but since they were performed on grains of pollen taken from the Shroud in the 1970s—shades of Frei-Sulzer—doubts about their veracity persist.

From its earliest days the Shroud has been a money spinner and, judging from the endless chain of books and articles that appear on the subject, its commercial appeal is undiluted. Nowadays most commentators attempt to concern themselves with the Shroud's "lost years," those thirteen centuries between the crucifixion and its mysterious appearance in a French backwater, with some claiming to have charted the Shroud's progress from the sepulcher in Jerusalem to Edessa in Mesopotamia, before it disappeared in the 1204 sack of Constantinople, all the way to Lirey in France. Sadly, the evidence they provide is of the kind that customarily accompanies breathless accounts of extraterrestrial architects, submerged civilizations, and landing strips on the Andes.

So if the Shroud is a forgery, how was it made?

No one has yet provided a satisfactory answer. It is inconceivable that merely pressing the cloth against a dead body would have produced such a perfect image; problems of scale and distortion would be immediately apparent as the two-dimensional cloth was peeled off a three-dimensional body. As Dr. Michael Baden, the former chief medical examiner for New York City, has noted, the image is "too good to be true . . . human beings don't make this kind of pattern."[22]

Which leaves only some kind of painting? Or a variant of brass-rubbing? Or could it really have been caused, as some would have us believe, by vapors emanating from a crucified look-alike, sacrificed to manufacture a relic?[23]

Or is it just an outrageous fluke, the unintended by-product of fiendishly clever trickery and degrading pigments, that happens to manifest itself in a photographic negative? It should not be forgotten that it was modern photography gave the Turin Shroud its current superrelic status. Without the camera, this strip of linen would be a lightly regarded curiosity, nothing more; though quite why it would be necessary to anticipate the invention of photography to share this miracle with the world is, frankly, beyond comprehension.

Whatever awe is accorded to the Shroud should be directed, instead, at its originator. More than six hundred years ago an unknown artist went into his studio and created a forgery of such subtlety and skill that it has fooled generations. Modern arrogance clings to the omnipotence of science, expecting it to solve all problems. When it fails to do so—as in this case— gloating enemies are quick to thumb their noses and trumpet the presence of paranormal intervention. This is just plain silly. History is littered with puzzles and artifacts that defy modern analysis—the pyramids at Teotihuacán in Mexico; Stonehenge; and the statues on Easter Island, for instance—and only the most fanciful would attribute any of these to mystical intercession.

Within years of its unheralded appearance in a tiny French church, the Shroud was condemned by the Vatican as a phony, and half a millennium later three independent teams of radiocarbon dating scientists reached the same conclusion.

Brilliant in its conception, magnificent in its execution, the Turin Shroud was a fraud in 1355, and the fraud continues, with no solution in sight. In that sense, it truly is the perfect crime.

Chapter 2

Napoleon Bonaparte (1821)

Poison, Poison Everywhere

reat heroic leaders are not supposed to die accidentally. Had Napoleon Bonaparte, France's supreme general and arguably the finest military commander of modern times, taken a musket ball in the chest or been eviscerated by a saber thrust in battle, then the gods would have been assuaged and the academics could have busied themselves assessing his role in history. Unfortunately, this potbellied little Frenchman wasn't so accommodating. He died in bed, and at a relatively early age. As invariably happens when notable figures expire out of turn, it didn't take long for the rumor mill to start grinding, with most of the accusing whispers hinting that perfidious Albion had been up to her old tricks again. Napoleon would have been delighted. All his military life he'd been a thorn in the side of the British, and he wasn't about to cut them any slack in death.

Five and a half years of stultifying exile on the island of St. Helena had turned the fifty-one-year-old battlefield maestro into a bloated, careworn wreck. Even so, when he did eventually die, on May 5, 1821, the circumstances were sufficiently unusual to merit an autopsy. Because its results were so ambiguous, so confusing, it left many of his countrymen in no doubt at all: France's greatest hero *had* been secretly murdered by his British captors.

It's an intriguing possibility.

Try to picture Alcatraz dumped in the middle of the South Atlantic and you get some idea of St. Helena when Napoleon stepped ashore on October 15, 1815. Except that on this particular prison island there was a garrison of three thousand troops and just one inmate. As far as the British were concerned, it was a case of once bitten, forever shy. Just seven months earlier, Napoleon had returned from exile on Elba, just off the coast of Italy, and sneaked back into France, ruling for the glorious Hundred Days until his crushing final defeat at Waterloo. The penalty for such impertinence was brutal: banishment to the outmost edges of civilization.

St. Helena is one of the remotest spots on Earth, a flyspeck of volcanic rock peeking above the mountainous waves of the South Atlantic. Twelve hundred miles to the east lies the coastline of Africa: half as far again, this time in a westerly direction, and you wash up on the shores of Brazil. The island was hot, unsanitary, riddled with disease, and above all boring. For a restless genius such as Napoleon, who had once held the destiny of Europe in his hands, such impotent tedium was ignominious beyond belief.

Longwood House, which would be home and prison to the ex-emperor and his retinue of retainers and former officers, was situated toward the middle of the island, up in the hills. The house matched the climate, damp and inhospitable, just as the British governor, Sir Hudson Lowe, intended. Lowe was not a pleasant man. His avowed mission was to make Napoleon's life as uncomfortable and humiliating as possible. Idle retirement was out of the question. Day and night sentries paced noisily outside Napoleon's quarters, driving him to paroxysms of rage, while he was taunted by rare glimpses, through the mists, of the convoy of ships that constantly patrolled the shoreline, a bitter reminder that any rescue attempt was doomed to failure.

Gradually the relentless psychological warfare took its toll, and Napoleon's health went into long and steady decline. From May 1816 he was attended by Dr. Barry O'Meara for a string of ailments, including insomnia, headaches and gout, all of which sapped his already flagging morale and added to his melancholic irritability. He fluctuated between bouts of lassitude and manic bursts of hyperactivity until September 1817, when the symptoms became more marked and he started to complain of pain on the right side of his abdomen. All witnesses reported significant swelling of the ankles and general weakness in his legs. O'Meara diagnosed hepatitis and administered calomel, a vicious purgative of the time made from mercury chloride.

As O'Meara grew ever closer to his patient—Napoleon enjoyed excellent relations with his entire British medical team—this provoked bilious accusations from Lowe that the doctor was deliberately scaremongering in

Napoleon in exile on St. Helena, guarded day and night by the British.

a bid to raise sympathy for Napoleon in Europe and secure his repatriation on health grounds. Lowe was consumed by self-interest. Any mention of hepatitis in official reports could only reflect badly on his command of an island that had become a hellhole (among the British garrison, the death rate in the ranks soared to 7 percent a year). Politics got the better of medicine, with the result that, in July 1818, Lowe succeeded in having O'Meara recalled to London.

There was no improvement in Napoleon's condition. On the night of January 16, 1819, he was struck down by a relapse so severe that many feared the worst. Another physician, Dr. John Stokoe, brought in to treat the emperor, confirmed O'Meara's diagnosis of hepatitis, though this was disputed by other doctors present, fueling Lowe's paranoid suspicion that Napoleon was in all probability faking the illness.

Support for Stokoe came from yet another doctor, Francesco Antommarchi—like Napoleon, a Corsican—who arrived at Longwood in September 1819. He was convinced that Napoleon was suffering from liver disease. His recommendation that the sluggish Napoleon should take more physical exercise at first produced a marked improvement; however, he was not in remission for long, and by the middle of 1820 the illness renewed its insidious creep. A crippling pain stole up the right side of his abdomen, into his shoulder, bringing with it fevers, coughs, chills in his legs, gingivitis, and alternating diarrhea and constipation.

The slide accelerated at an alarming rate. In March 1821 he was confined to his bed as his symptoms worsened, his doctors powerless to help him.

Early-nineteenth-century medicine was rugged stuff; even if the ailment didn't kill you, chances were the "cure" would. High on the list of any physician's remedies was purging, for it was felt that only by expelling all toxins could the body heal itself. It sounded fine in theory, but the practical application was often dangerously debilitating. Here, Antommarchi, whom Napoleon loathed and sneeringly derided for his perceived incompetence, prescribed tartar emetic (a vomiting agent containing antimony), which further weakened the patient. When an English naval doctor named Archibald Arnott was then summoned, at first he didn't consider the illness to be that serious, but ultimately he agreed that Napoleon was nearing the end. After consulting with two British colleagues, and over Antommarchi's vehement opposition, at 5:30 P.M. on May 3, 1821, Arnott prescribed a massive dose of calomel, ten grains compared with the normal one or two. Napoleon's response was to begin vomiting blood; then he lapsed into a coma from which he never recovered. Forty-eight hours later, he was dead.

Crowded Autopsy

The next day his body was laid out on the billiard table at Longwood House, awaiting autopsy. A seven-man team of British doctors watched gimlet-eyed as Antommarchi opened up the corpse. Afterward, although no one seemed in complete agreement as to exactly what had been found— Antommarchi alone produced two markedly conflicting reports—a reluctant consensus did emerge. As suspected, the liver was enlarged and there was a large ulcer in the stomach, which appeared to have caused the severe hemorrhaging that marked Napoleon's final hours. Near the pylorus, where the stomach empties into the duodenum, they found a swelling, which, after much heated discussion, was classified as a cancerous growth. As Napoleon's father and sister had both died from a similar ailment, the doctors eagerly grasped this opportunity to state that their patient had succumbed to a hereditary predisposition.

So it was official, then: Napoleon died of cancer.

Few in France were convinced; especially when the contents of Napoleon's will were made public. In it, the exiled emperor had written: "I am dying before my time, murdered by the English oligarchy and *its hired assassins* [emphasis added]."[1]

Most level-headed interpretations assumed this to mean that Napoleon was referring figuratively to the war of nerves waged by the hated Hudson Lowe; for a noisy minority, however, there was another, far more sinister interpretation: the emperor's end had been hastened by the deadly hand of poison.

Among trained historians the poisoning theory has always been lightly regarded, just one of those frivolous interludes that crop up every now and then and disrupt serious academic study, but in the mid-1950s a Swedish dentist, Sten Forshufvud, donned the mantle of a Scandinavian Sherlock Holmes and set out to crack the case of "Napoleon—Murder Victim?" An unabashed Napoleon idolater—his house in Göteborg was filled with portraits, busts, and statues of the emperor—Forshufvud adopted the methodology favored by most conspiracists: first start with a conclusion, then work backward, searching for scraps of "evidence" to prop up the hypothesis.

He immersed himself in several contemporary accounts of Napoleon's illness, and found what he wanted in a diary kept by the emperor's valet, Louis Marchand. A highly subjective reading of this document satisfied Forshufvud that, prior to his death, Napoleon had exhibited clear signs of chronic arsenical poisoning.

Throughout history arsenic has been the prince of poisons. Called "inheritance powder" for the way in which it has shaped genealogies and family fortunes, it is almost entirely tasteless, odorless, and wonderfully portable. With its ability to replicate the symptoms of dysentery, food poisoning, cholera, and countless other common ailments, arsenic was the toxin of choice for Europe's professional poisoners, a loosely connected gaggle of talented psychopaths who hired out their unique skills in the seventeenth century to settle disputes among the aristocratic and the affluent.

Forshufvud was convinced that Napoleon had been the victim of just such a poisoner. After all, when his body had been exhumed in 1840 for reburial in France, hadn't everyone commented on its remarkable state of preservation, often a strong indicator of the presence of arsenic? But Forshufvud needed proof. Denied access to Napoleon's body—it lies in a shrine at the Dôme des Invalides in Paris, visited by millions each year, and no French government would dare permit such desecration—he was forced to look elsewhere. His golden opportunity came when he took possession of a single strand of hair, snipped from Napoleon's head on the day of his death by the ever-faithful Marchand.

Because arsenic gets into hair via sweat and other sebaceous secretions and binds strongly to the keratin molecules, it leaves an identifiable record of contamination. If Napoleon did ingest arsenic, it was odds on that this hair would be able to provide confirmation. Forshufvud sent the hair to Professor Hamilton Smith at the Glasgow University Forensic Science Laboratory, where it was subjected to neutron activation analysis (NAA).

At the time of this experiment—1960—NAA was cutting-edge science in the field of trace analysis. It works by placing the sample in a capsule, which is then inserted into a nuclear reactor and bombarded with neutrons to make it radioactive. By measuring the rate at which radioactive atoms disintegrate, it is possible to identify that sample's trace elements. Any trace of arsenic would be expressed in parts per million (ppm).

In Glasgow, Smith weighed the lone hair strand and sealed it in a polyethylene container. Then the hair strand and a standard arsenic solution were both irradiated for twenty-four hours. The results were startling: the hair contained 10 ppm (parts per million) of arsenic, almost thirteen times what was then considered to be the normal level of 0.8 ppm.

Excited though he was by this discovery, Forshufvud kept his feet on the ground. Just because the hair showed traces of arsenic, that didn't amount to proof of deliberate poisoning. Because arsenic also can bind itself to the hair through external contact, such as the earth in which a coffin has lain, Forshufvud wondered if there was any way to prove that the arsenic had been taken internally.

Smith had the answer: section analysis. "If arsenic was absorbed from the natural environment," he told Forshufvud, "the analyses of the hair would show a constant amount of arsenic along its length. If, on the other hand, arsenic was ingested into the body at intervals, the hair strand would show peaks and valleys of arsenic in each section."[2] Furthermore, because hair grows at approximately 0.014 inch per day, Smith thought he could calculate the time between the peaks.

He performed more than a hundred tests on another three-inch sample of Napoleon's hair, this time clipped by one of his valets, Jean-Abraham Noverraz. Broadly speaking, this hair represented a six-month log of

Napoleon's life. Section analysis indicated that the arsenic had not come from the environment, because its content was not constant. And the ppm numbers defied belief—ranging from a low of 2.8 ppm to a high of *51.2 ppm!*

Here, Forshufvud believed, was the Holy Grail, irrefutable proof that Napoleon had been deliberately poisoned over an extended period of time.

In 1961 he published his findings and caused an uproar, especially in France. This wasn't some wild-eyed polemic cobbled together to make a quick buck; this was a closely reasoned argument that Napoleon had, indeed, been murdered, backed up with what appeared to be solid scientific evidence. Furthermore, because the team had been able to locate and test another hair dating from early 1815, and this, too, showed abnormal levels of arsenic, the conclusion was ineluctable: someone had been trying to poison Napoleon long before he fell into the hands of the British.

In 1974 Forshufvud joined forces with Ben Weider, a like-minded Canadian multimillionaire who'd made his fortune in the physical fitness industry, and together they set about spreading their revisionist gospel. Weider brought to the project marketing smarts and a sharper focus on the clinical pathology of Napoleon's death. Even without the evidence of the NAA results, he had already eliminated cancer as a cause of death because the autopsy recorded that Napoleon was obese, and terminal cancer tends to be a wasting disease. Their joint book, *Assassination at St. Helena,** asserted that, according to eyewitness statements, Napoleon displayed thirty of the thirty-four known symptoms of arsenical poisoning.

Moreover, they even claimed to have identified the culprit who administered the poison: Comte Charles-Tristan de Montholon, an officer who had served with Napoleon since 1806. As the *sommelier* at Longwood House, Montholon was, so the reasoning went, able to poison Napoleon with his own personal wine, which came from South Africa.

By working out timelines correlating the results of the section analysis to Napoleon's recorded day-to-day activities, Forshufvud and Weider concluded that only Montholon had access to Napoleon on every occasion when his hair showed an upward spike in arsenic contamination. As their theory took root, Montholon was painted in ever darker hues. He was portrayed as a royalist agent, dispatched by the Bourbons to guarantee that Napoleon never again set foot in France, a lascivious playboy officer plagued by deep and besetting money problems. As an example of his profligacy, they pointed out that, although a major beneficiary under Napoleon's will,

---

*Vancouver, BC, Canada: Mitchell Press, 1978.

inheriting more than two million francs, a huge sum in those days, Montholon still wound up bankrupt in 1829, having to flee to Belgium to escape his creditors. As evidence of bad money management, it was compelling; as proof of murder, it didn't amount to a row of beans.

Nevertheless, the book went on to claim that Montholon actually started poisoning Napoleon in 1816, gradually wearing down his resistance with repeated doses of arsenic, until he deteriorated to the point where no one would notice when he was actually dispatched by some other toxin. Weider termed it "[poisoning] in the classical manner of the nineteenth century."[3] Apparently this called for a long "cosmetic phase"[4] to allay everyone's suspicions, followed by a "lethal phase,"[5] in which a quite different poison was used to administer the *coup de grâce*.

The deadly agent in this instance, according to Weider, was hydrogen cyanide. The assassination plan worked like this: After years of being dosed with arsenic, in March 1821 Napoleon was subjected to a punishing regime of tartar emetic, designed to destroy his stomach lining. Then, on April 22, he was given a drink called orgeat to quench his raging thirst. Mixed with this drink were ground-up peach stones, which contain traces of cyanide and reacted with the colossal dose of calomel (mercury chloride) on May 3 to form a lethal cocktail that finished off Napoleon.

This, then, was the theory, and it proved very lucrative for its proponents. Sensational accounts of Bonaparte's "murder" sold in the millions— so much so that, for many, the theory became fact, proof positive that Napoleon had been deliberately poisoned.

Then, as so often happens, doubts began to surface.

Poison, Poison Everywhere . . .

Arsenic is the twentieth most abundant element in the Earth's crust. All of us have traces of it in our bodies and our hair. As to what constitutes a "normal" level of hair contamination, the scientists are in dispute, but the 0.8 ppm quoted by Forshufvud and Weider seems to be unfeasibly low. Environment plays a crucial factor. In heavily polluted Mexico City, for instance, readings of 4 ppm are not uncommon, while across the Atlantic, in metropolitan Glasgow, the average is 3 ppm, although one perfectly healthy student did register an arsenic level of *12 ppm*. Such divergences, as huge as they are baffling, have led one American laboratory that specializes in NAA to opine that anything less than 10 ppm might be considered "normal."[6]

Also, hair analysis for arsenic is a very unreliable indicator of serum arsenic levels when a specific individual is tested in isolation. Only by comparison with a range of reference values from a group of contemporaries would it be possible to obtain scientifically sound results, and in the absence of hair samples from other residents of St. Helena in 1821, we have no way of knowing what the average concentration of arsenic really was at that time. Having said that, even though environmental factors might conceivably account for the 10 ppm of arsenic found in the first Napoleon hair analyzed, clearly they don't begin to explain the astronomical levels found by section analysis in the second specimen. Which leads us to the next question: How accurate was NAA in 1960?

Although NAA may appear straightforward, it was, in this case, extremely difficult to apply because of the hair's relatively low mass and the risks of the sample being polluted during handling. In 1960 the technique was less sophisticated than it is today, which made isolating some elements such as bromine, arsenic, and antimony a fiendishly tricky process.

And then there were those "timelines," the smoking-gun chronology that pinned the stigma of "poisoner" onto Montholon. Analysis carried out at the Saclay Nuclear Research Center in France has shown that using hair segments as short as those tested by Smith produces an accuracy level of plus or minus 20 percent, enough to make nonsense of any attempt to establish when Napoleon was poisoned. These discrepancies occur because arsenic, rather than remaining at the point where the hair has grown, sometimes spreads throughout the strand by capillarity.

More recently another of Napoleon's hairs was tested by the University of Toronto's "Slowpoke" reactor and analyzed for arsenic, bromine, and antimony, among other things. The hair showed only 1.5 ppm of arsenic, way below the level necessary to indicate chronic poisoning. The hair did, however, have 6 ppm of antimony, which may have given a false indication of high levels of arsenic in 1960. This is important, since the use of antimony was common in medical practice in the nineteenth century.

In fairness to Weider it should be noted that a 1995 FBI examination of yet more of those Napoleon hairs—Napoleon's onetime aide Comte Flahaut, remarked caustically in 1862, "I have seen so many locks of hair purporting to be that of the emperor over the last twenty years, I could have carpeted Versailles with it!"[7]—disclosed readings of 33.3 ppm and 16.8 ppm.

If all these hairs were genuine, and this is by no means certain, then arsenic clearly got into Napoleon's system somehow. But did this amount to proof of willful poisoning?

In 1996 Dr. Philip Corso of Yale University Medical School and Dr. Thomas Hindmarsh of Ottawa University published a paper that brought some much-needed objectivity to this overheated field. "As every forensic scientist knows, the diagnosis of chronic arsenic poisoning cannot be made upon elevated arsenic concentrations in hair alone. . . . Before a diagnosis of chronic arsenic poisoning can be made, the characteristic clinical features must be present as well."[8]

Earlier, Weider, and Forshufvud had stated that eyewitnesses noted no fewer than thirty of the known symptoms of chronic arsenic poisoning in the dying emperor. These included lassitude, chills, stomach pains, insomnia, alternating diarrhea and constipation, vomiting, raging thirst, and itching of the skin—the list goes on. But as we have seen, arsenic's nefarious popularity resides in its ability to mimic the symptoms of other, totally innocuous illnesses. In such a situation, broad indicators are meaningless; what one needs are the more specific symptoms of arsenical poisoning.

And these, according to Corso and Hindmarsh, were conspicuously absent. Where was the constant raindrop pigmentation of the skin, particularly around the armpits, groin, temple, eyes, neck, and nipples, sometimes spreading over the chest and shoulders? Nor was there any sign of hyperkeratosis, a marked thickening of the skin on the palms of the hands and soles of the feet. Indeed, at the autopsy, Napoleon's skin was reported as "white and delicate, as were the hands and arms,"[9] a far cry from what is normally found in victims of chronic arsenical poisoning.

Another peculiar phenomenon associated with arsenic intoxication is Mees's lines, transverse bands of white and dark lines (usually from one to three) on the nails of the fingers and toes. They are caused by precipitations of arsenic within the keratin matrix of the fingernails, and are first noticed approximately eleven days after the onset of contamination and remain as a succession of bands for more than three months. The significance of prominent Mees's lines is now well established as a sign of arsenical poisoning, leading one scientist to state that they represent "the most helpful diagnostic finding of arsenic polyneuropathy,"[10] with about 80 percent of cases exhibiting this symptom.

Not one doctor attending Napoleon's autopsy noted any evidence of Mees's lines.

And then there was the baffling question of Napoleon's obesity. When Weider claimed this as strong evidence that cancer had not killed Napoleon, few would have argued the point. However, Weider and Forshufvud went farther. In their text, obesity miraculously transmogrifies to became an undisputed symptom of chronic arsenical poisoning. This flies in the face of received medical wisdom, which sides with Hindmarsh's view that "weight *loss*"[11] is one of the commonest symptoms of arsenical poisoning.

So, of the four classic symptoms of chronic arsenical poisoning, not one was found to be present during Napoleon's autopsy.

Even setting aside the clinical objections, what about the logistics of murdering Napoleon in the manner suggested? The biggest single drawback is the time scale. According to the conspiracists, the systematic poisoning of Napoleon began even before Waterloo—only such devious intervention could account for the great man's curious battlefield detachment, apparently—which means that the poisoner's lethal activities were drawn out over six years at least. If true, this must be the slowest assassination in history. And for what reason?

Weider's claim that the poisoner was acting in the "classical manner of the nineteenth century" disregards the fact that by the time of Napoleon's death the "professional poisoners" of Europe had been extinct for almost two hundred years. Also, there is no credible evidence to suggest that they ever adopted a two-tier approach to their murderous work. Softening up a victim for years with arsenic, producing a set of symptoms that becomes synonymous with that person, only to then switch poisons for the final deadly dose harks of lunacy. Such obvious amateurishness would have virtually guaranteed a trip to the headsman's ax.

Cyanide kills in a very different way to arsenic. It prevents oxygen from being distributed through the body, leading to suffocation. Above all, it is lightning fast. The first symptoms of cyanide poisoning are rapid heartbeat, headache, and drowsiness, followed by coma, convulsions, and death. At this point the victim's face is usually bright red as a result of the change in hemoglobin mentioned above. According to medical examiner Dr. Michael Baden, "[cyanide] turns the blood and skin scarlet."[12] It is inconceivable that such garish symptoms could have gone unrecorded by the doctors attending Napoleon. Yet none was observed.

Nor was any evidence of poisoning found during the autopsy, not that this should surprise anyone. Arsenic is invisible on the dissecting table, whereas cyanide occasionally leaves just the faintest odor of almonds, though, interestingly, not everyone can smell it. Also, in 1821 toxicological detection was still in its infancy. A mere 30 years had passed since a chemist named Johann Metzger first discovered how to identify if food contained arsenic, through what became known as an "arsenic mirror." A significant refinement to this technique came in 1806, when Dr. Valentine Rose of the Berlin Medical Faculty designed a test that, for the first time, permitted the detection of arsenic in the human body.

Although rudimentary and unable to detect minute quantities of arsenic, Rose's test was considered state of the art at the time of Napoleon's death, and the reason it wasn't employed was because no one attending the autopsy had any shred of evidence to suspect that Napoleon had been poisoned.

Weider continues funding expensive tests on still more strands of hair allegedly taken from Napoleon's body. In June 2001 a team of French scientists led by Dr. Pascal Kintz, deputy director of the Strasbourg Institut de Médécine Légale, reported finding "very large concentrations of arsenic" in samples provided by Weider. Kintz did, however, inject an admirable note of caution, saying, "We can't know the origin of the arsenic, nor be one hundred percent certain that the locks in question are those of Napoleon."[13]

Which still, of course, leaves one nagging question: Where did the arsenic in his hair come from?

Numerous theories have been advanced. None seems wholly convincing. Hindmarsh has his suspicions. "My only explanation is that arsenic was used as a preservative for hair sold as mementos,"[14] he said. In the early nineteenth century, arsenic was terrifyingly ubiquitous. It could be found in medicines, tonics, as a whitener for ladies' complexions, in flypapers, as a dusting agent for wigs, even in wallpaper, and, in minute doses, it was even used recreationally as a mind-altering drug that some said acted as an aphrodisiac. Such availability meant that arsenic could have gotten into Napoleon's hair by any number of means, all innocuous.

So if there is not a scintilla of evidence to suggest willful poisoning, and we accept Weider and Forshufvud's assertion that Napoleon did not succumb to cancer, what exactly did kill the emperor? Hindmarsh points the finger of blame squarely at that huge dose of calomel on May 3, 1821. "I think this probably precipitated a bleed from the cancer. It's unusual for a cancer of the stomach to bleed unless it's irritated in some way, and it's apparent from the autopsy that he did suffer bleeding from the bowel and vomiting of blood."[15]

No other solution fits the facts so neatly. So maybe those skeptical Frenchmen were correct all along: perhaps Napoleon *was* killed by the British—not by poison, but by bad medicine.

Chapter 3

Alfred Packer (1874)

The Colorado Cannibal

D igging up bodies has become an America growth industry. Uneasy lie the bones of the famous or notorious who die in anything other than the most mundane circumstances; chances are, it won't be long until the forensic resurrectionists congregate at graveside, with their shovels, pickaxes, arc lights, video cameras, and all the other paraphernalia of the well-funded exhumation. The often lucrative world of vanity forensics can take many forms; always, though, the declared intent is the same—to finally uncover the absolute truth.

If only it were that simple. For as the following case demonstrates, the truth can often be as elusive in death as it is in life.

Throughout most of the nineteeth century, Alfred Packer has either been damned as a cannibal or regarded as a source of thigh-slapping merriment. For his pains, this self-confessed man-eater has become a stop on the Colorado tourist trail, the recipient of his very own "day" at the University of Colorado, a face on a T-shirt, the subject of a Phil Ochs's folk song, a cult movie musical, and the launch pad for a thousand cringe-making puns. Out where Packer starved and carved in western Colorado, institutions were quick to cash in on his notoriety. One, the "Cannibal Café," proudly boasted, "We'd Love to Have You for Dinner"; another diner offered the "Packer Platter—for man-sized appetites;" while in Lake City the

Alferd Packer Society* could always raise a guffaw with its motto "Serving our fellow man since 1874."

Let's get one thing straight at the outset: Packer was not, as has often been stated, "the only man in U.S. history ever convicted of cannibalism,"[1] because cannibalism has never been a crime in most states (oddly enough, only Idaho has seen fit to put a law criminalizing anthropophagy in the statute book). No, this Pennsylvania-born shoemaker turned mountain guide was ultimately convicted of nothing more heinous than manslaughter. The question is: Should Alfred Packer have been convicted of any crime at all?

It's the fall of 1873, and Packer, a thirty-year-old army reject mustered out of the Eighth Regiment, Iowa Cavalry, because of epilepsy, has grubbed and hustled his way to Provo, Utah. While scuffling around for work, he got news that a twenty-strong party of prospectors bound for the Rocky Mountain gold fields was on the lookout for an experienced guide. Despite his unfamiliarity with the region, the squeaky-voiced Packer piped up with an offer of his services. Calamitously for all concerned, he got the job.

The winter that year was frigid, full of blizzards, killing winds, and stupendous snowdrifts that seemed to reach to the leaden skies, and had the group not found sanctuary at the Southern Ute camp of Chief Ouray in the Uncompahgre Valley, all would have perished.

One would imagine that such a close brush with death would have inspired significantly enhanced levels of self-preservation, but greed got the upper hand of common sense, and against the advice of Chief Ouray, who warned of even worse weather to come, on February 9, 1874, Packer and five others decided to strike out for Gunnison, Colorado. As forecast, a colossal snowstorm swept in and blanketed the mountains to a depth of many feet. Lakes froze hard, paths and tracks disappeared, and for weeks nothing was heard of the breakaway group until, on April 16, alone and toilworn, Packer staggered into the Los Pinos Indian Agency near Saguache, Colorado.

The story he told—of having been separated from his companions in a blizzard—sounded credible enough, but something about his appearance just didn't ring true. Although undeniably exhausted, Packer looked remarkably well nourished for someone who had spent two starving months in the mountains. And how had he laid his hands on all that money he was

---

*Although his legal name was Alfred, Packer had a tattoo that read "Alferd" and invariably signed his name that way. On the document that matters most, his parole, the name is given as Alfred.

now splurging on card games and liquor? And that Wells, Fargo bank draft made out to one of the missing men? When he got drunk—which was often—saloon bar cronies noticed how certain discrepancies began to creep into his story. Just a slip here and there, but enough to eventually reach the ear of General Charles Adams, the Los Pinos agent, who angrily confronted Packer and demanded that he lead a search party in an effort to locate the missing men.

They set out on May 1. Packer proved to be a reluctant guide, curiously hazy and forgetful. After three days spent meandering through the mountains, along gulches, and across passes still covered with snow, he couldn't maintain the pretense any longer. In a rush of conscience he unburdened himself to Adams.

The six-man party had become marooned in the San Juan Mountains, he admitted, snowbound and driven half mad by hunger, reduced to living on roots and snails, even chewing their moccasins for sustenance. Realizing that the group faced certain extinction if they stayed put, Packer

Alfred Packer, the "Colorado cannibal," as he looked in 1883, when jailed for forty years. (Courtesy of Colorado State Archives)

volunteered to search for a safe route off the mountains. For days he battled the blinding snow. In the end, whipped by the nonstop blizzard and the subzero temperatures, he was forced to abandon his plan. More through luck than geographical smarts, he somehow fumbled his way back to camp.

A ghoulish sight greeted him. Israel Swan, the oldest member of the party, who had died from exposure and starvation, was being cut up and eaten by the survivors. Without a second thought, Packer reached for his knife and joined in the feast. As the weather worsened, one by one the other party members succumbed—first James Humphrey, then Frank Miller, and George Noon. All were eaten.

This only left Shannon Wilson Bell, a demonic red-haired prospector, and Packer. Tension levels between them escalated daily until, so Packer claimed, Bell snapped and flew at him in a homicidal rage. To defend himself, Packer had grabbed a gun and shot his demented partner, then set about butchering the body for food.

Packer concluded his confession with an admission that he had pocketed his fellow travelers' money.

Adams thought hard. Every God-fearing instinct in his body told him that Packer must be guilty of something. But what? Utterly perplexed, the general requested clarification from Washington. In the meantime, he placed Packer under arrest. As the weeks dragged by with no news from the capital, Packer's edginess increased to the point where he decided to take matters into his own hands. On August 8 he mysteriously escaped from custody in Saguache County, leaving behind him strong rumors of jailhouse bribery, and vanished.

Which was just as well, really, because two weeks later, five male bodies—one headless—were discovered by a bluff on the shores of Lake San Cristobal. Although the carcasses had been stripped of large chunks of flesh, there was enough remaining to indicate that the men had been bludgeoned to death. Immediately an arrest warrant was issued, charging Packer with multiple counts of murder.

Almost nine years would pass before the law caught up with Packer. He was caught by one of those unaccountable coincidences that so often characterize crime detection. One of the original prospectors on that ill-fated trip, a man named Frenchy Carbazon, happened to be drinking in a saloon in Fort Fetterman, Wyoming, when he heard a familiar high-pitched cackle of laughter. A stealthy glance in the reveler's direction confirmed his suspicion; there was no disguising that runty face. Three days later, on March 14, 1883, John Schwartze was arrested at Wagonhound Creek, some thirty miles from Fort Fetterman, and charged with being the wanted fugitive Alfred Packer.

The *Saguache Chronicle* gleefully recorded the arrest. "After Nine Years a Fugitive From Justice, the Capture is Effected of the Human Ghoul who Murdered and Grew Corpulent on the Flesh of His Comrades,"[2] ran the headline.

Within days of his arrest, Packer made yet another confession to Charles Adams, and it was this semiliterate account that would form the bedrock of his defense when he was extradited to Colorado, and the Hinsdale County Courthouse at Lake City, to stand trial on April 6, 1883, for the murder of Israel Swan.

Packer insisted that he had returned to the camp to find "the redheaded man [Bell] who acted crazy in the morning sitting near the fire roasting a piece of meat which he had cut out of the leg of the German butcher [Miller]."[3] Nearby lay the lifeless bodies of the other three prospectors. "They were cut in the forehead with the hatchet, some had two some three cuts . . . when the man [Bell] saw me, he got up with his hatchet towards me when I shot him sideways through the belly, he fell on his face, the hatchet fell forwards. I grabbed it and hit him in the top of the head."[4]

Admits to Cannibalism

Afterward, Packer claimed, he had hunkered down, building a rough shelter to see him through the rest of winter, all the while feeding on the bodies of his former companions. When the snow at long last cleared, he had trudged on to Los Pinos. In this confession he made no attempt to deny his cannibalism but insisted that the only person he had killed was Bell, in self-defense. It was a story neither the jury nor Judge Melville B. Gerry found credible, and Packer was found guilty of murder.

Legend has it that Judge Gerry sentenced Packer as follows: "There was seven Democrats in Hinsdale County, but you, you voracious, man-eatin' son of a bitch, you ate five of them. I sentence you to be hanged . . . as a warning against reducing the Democratic population of the state." [5]

In reality, Gerry's sentencing speech was less politically charged but just as florid, mounting to a dramatic crescendo as he condemned Packer to "be hung by the neck until you are *dead, dead, dead!*"[6]

Overnight Packer had become the most hated man in Colorado, and there were genuine fears that a lynch mob might attempt to anticipate due process of law while the sentence was being appealed. To ward off this threat of outside intervention, the authorities transferred Packer to Gunnison, and his execution was stayed.

Packer's notoriety is hard to explain, since cannibalism was not uncommon in the brutally harsh mountain winters that faced transcontinental

American explorers. The Donner-Reed tragedy of 1846–1847 was merely the best known. On that occasion, not only were dead companions consumed, but also two Indian guides, who had refused to eat human flesh, were shot and eaten—a deed as reprehensible as anything Packer was ever accused of. Yet there were no arrests, no trials, scarcely any condemnation.

All of which must have been cold comfort to Packer as he awaited his date with the gallows.

Then came a sensational development. An astute lawyer had done some research and discovered that when the crime occurred, Colorado was still a territory—it didn't achieve statehood until August 1, 1876—and that the state legislature had neglected to reenact the territorial murder law. This meant that murders committed during a certain time frame were *not illegal*. It sounded incredible, but the Colorado Supreme Court had no other choice: On October 30, 1885, they duly overturned Packer's murder conviction.

Further scouring of the statute book brought a huge sigh of relief from the Colorado judiciary; it transpired that the manslaughter laws for the contested period *had* been properly enacted, which allowed the state to charge Packer with five counts of manslaughter. His trial ended on August 5, 1886, with him being sentenced to forty years' imprisonment in the state penitentiary at Cañon City.

A model prisoner, while behind bars he became an unlikely *cause célèbre*, with a string of attorneys applying for retrials and pardons. Although many of these motions had undeniable legal merit, it wasn't until Polly Pry, a "sob sister" reporter for the *Denver Post,* threw her lachrymose talents behind Prisoner 1389 at about the turn of the century that the "Free Packer" campaign really took hold. Pry was unbeatable when it came to tugging heartstrings, and she took dead aim at Governor Charles Thomas. The nonstop badgering paid off. On January 1, 1901, as his last official act, the outgoing governor granted a parole for Colorado's most notorious prison inmate.

After his release Alfred Packer lived quietly on the outskirts of Denver, working as an occasional janitor at the *Post,* until his death on April 23, 1907. Each year thousands of tourists visit his simple grave at Littleton, all eager for a glimpse of the last resting place of the "Colorado Cannibal."

But was Alfred Packer really the blackhearted demon of mountain lore, or was he, as Pry insisted, the tragic victim of a justice system that overlooked facts in favor of emotion?

The quarrel raged through most of the twentieth century. In 1989 a battalion of forensic scientists converged on Lake City, Colorado, determined to settle the argument once and for all.

Alfred Packer just before his release in 1901. (Courtesy of Colorado State Archives)

Many people are uncomfortable with the notion of exhumation. Even when there is a clear-cut legal need, the act of unearthing a corpse can arouse deeply held feelings of unease. Ethical dilemmas figure large in this debate, centering on the public's right to know versus the dead's right to dignified rest. In recent years the number of what might be termed "historical exhumations" has grown exponentially. In this respect forensic science has been a victim of its own success. Because modern analytical techniques are so illuminating, the temptation to employ them indiscriminately has, at times, been hard to resist.

The scenario follows a predictable pattern: Someone with an agenda—usually either ego-driven or financial—gets a bee in his or her bonnet about some noted figure, applies for an exhumation order, and the next thing you know the excavators are on the scene and everyone is shoveling like a locomotive engineer. One of America's foremost anthropologists, Clyde Snow, finds the trend disquieting. Exhumations, he says, should be performed only if reputable historians feel it could shed light on critical historical issues. "I don't know that just because somebody out there has some doubts about what happened, that we should jump in and dig people up."[7]

Such skepticism cuts little ice with James Starrs, professor of law and forensics at George Washington University in Washington, D.C., and an archapostle of forensic resurrectionism. Over the years Starrs has turned his flamboyant attentions to a string of high-profile cases, among them Jesse James, Meriwether Lewis, Carl Weiss (the assassin of Huey Long), and Lizzie Borden. Overzealousness does seem to be a problem. (Taking issue with the heart attack that accounted for seventy-seven-year-old J. Edgar Hoover's demise is just the most notorious example.) The disappointingly inconclusive results that most of Starrs's efforts have produced thus far do tend to support the notion that he is more adept at generating headlines than epoch-changing revelations, but all bandwagons have to start somewhere, and for Starrs it was the "Colorado Cannibal" that first thrust him into the limelight from which he has so rarely strayed.

Starrs was convinced that the answer to what really happened in that dreadful winter of 1874 lay in a reexamination of the alleged murder victims. He told reporters, "From the bones we will be able to see bullet holes and hatch marks from where the skinning knife might have scraped the bones. We'll be able to tell if the victims were cannibalized, if they were struck by a hatchet blow, if they were really near to starvation."[8]

In the summer of 1989 Starrs assembled his grandiosely titled "Packer Project," a thirteen-strong team of archaeologists, pathologists, anthropologists, and technicians, and set out for western Colorado. The first problem was locating the bodies of the victims. Starrs declared he was "90 percent sure"[9] the bodies lay under a grave marker erected in 1928 in Deadman's Gulch, which lies between Cannibalism Plateau and Round Top Mountain, about two miles south of Lake City.

To find the bodies, the team used ground-penetrating radar (GPR). This highly sophisticated locating device operates by transmitting pulses of ultrahigh-frequency radio waves into the ground through a transducer or an antenna. When the transmitted signal enters the ground, it contacts objects or subsurface strata with different electrical conductivities. Some of the GPR radar waves reflect off the object or interface, while the rest of the waves pass through to the next interface.

Once the reflected signals return to the antenna, they are received by a digital control unit, which registers the two-way travel time in nanoseconds and then amplifies the signals. The output signal voltage peaks are plotted on the ground-penetrating radar profile as different color bands by the digital control unit.

In this instance the five mutilated bodies were exactly where Starrs thought they would be. After exhumation, the corpses were taken to the Arizona State Museum at Tucson for detailed analysis.

Examination of the bones revealed clear signs of foul play. At least four of the victims had been bludgeoned to death, three by a hatchetlike instrument, the fourth by a rifle butt or something similar. (As noted earlier, one skull was missing entirely.) Marks on the arm bones had the look of defensive wounds, caused in all probability when the victims had attempted to fend off blows. Distinctive scrape marks strongly suggested that the bodies had been carefully butchered with a skinning knife.

Significantly—and at variance with Packer's account of having shot Bell in self-defense—none of the victims showed any indication of having suffered a gunshot wound. Although a hipbone on one of the skeletons did have a hole in it, this had, according to Starrs, probably been caused by a hungry coyote gnawing on the corpse after the spring thaw of 1874, a conclusion supported by contemporary reports that the human remains were incomplete and showed signs of having been scattered by animals.

On the basis of this rather perfunctory analysis, Starrs felt confident enough to announce his verdict. It came in October 1989. "Packer was as guilty as sin and his sins were all mortal ones,"[10] he announced in his trademark high-flown manner. As for Packer's claims of self-defense, Starrs snorted his derision. "It is as plain as a pikestaff that Packer was the one on the attack, not Bell."[11]

Starrs's bombast puzzled many, as there seemed to be glaring holes in his data. By his own admission he had no idea which body was which, and the paucity of physical descriptions of the victims had left his team groping in the dark. Even had the scientists been able to determine whether the wounds were inflicted by cutting left-handed or right-handed, this would have been of little use, since they didn't know who in the ill-fated party was left- or right-handed. Most important of all, Starrs could not say, with any degree of certainty, which set of bones belonged to Shannon Bell.

Nor could the fact that the bones displayed similar nicks from a skinning knife be presented as conclusive evidence of a single killer. Bones recovered from known acts of cannibalism from around the globe, from the American Southwest, to France, to England, all display broadly similar cut marks. There are only so many ways one can butcher a carcass, and it would be a brave scientist, indeed, who stood up in court and professed to be able to identify whose hand wielded a skinning knife. Either Packer or Bell could have wielded that knife. Or both could have.

Starrs scoffed at such objections, and after the bones were reinterred in a single coffin under the grave marker in Deadman's Gulch, he issued his final judgment. "This latest evidence convicts him [Packer] beyond a shadow of a doubt. Packer was having his flesh filets morning, noon and night, even though he could have lived by killing rabbits. Packer was a fiend, base, brutish . . . barbaric."[12]

Case proved? End of subject?

Of course not.

Forensic Sniping War Breaks Out

Even some of the scientists who had worked on the "Packer Project" felt that Starrs had overegged the pudding somewhat in making such extravagant claims. Physical anthropologist Walter H. Birkby of the Arizona State Museum, who had carried out the analysis of the bones, while agreeing that Starrs's version of events was consistent with the evidence, urged caution: "It could possibly be the right scenario, but scientifically we cannot substantiate it."[13] He added that although the evidence clearly showed the victims had indeed been murdered and cannibalized, there was no proof to identify the culprit other than the fact that the wounds were inconsistent with Packer's testimony. "We'll never know who did it based on any solid evidence. We're never going to know,"[14] said Birkby.

Others have gone much farther, convinced Packer has been gravely defamed by the courts and by history. Ervan F. Kushner, a retired judge, who researched and wrote a book about the case,* felt strongly enough about Packer's innocence to seek a posthumous pardon. While former governor Dick Lamm agreed that Kushner's representations did much to rehabilitate Packer's reputation, he felt they were insufficient to warrant the issuance of a pardon.

Evidence of a far more persuasive kind came via the forensic laboratory.

For several years David Bailey, the head curator at the Museum of Western Colorado, had long believed in Packer's innocence. Long hours poring over trial transcripts and, especially, reading letters written by Packer, convinced him that a gross miscarriage of justice had occurred. He couldn't reconcile the image of Packer the heartless murderer with Packer the avuncular ex-con who carved dollhouses, gave candy to kids, and loved nothing more than to putter in his well-tended garden. (Bailey seems to have overlooked the fact that the passage of three decades and a well-lined belly might have greatly sweetened Packer's disposition.)

After his release from prison, Packer was interviewed by the *Denver Post*, and Bailey formed the unshakable belief that the story he told the newspaper at this time and held fast to on his deathbed was essentially true:

"He [Bell] came runnin' at me with a hatchet. He had the only hatchet in camp. I could see that he was mad. He made a kind of grating noise. I ran back. I had a revolver. When I got to the snowdrift, I pulled my gun. He came on the run after me, and when I got to the deep snow I wheeled 'round as quickly as I could and fired."[15]

Alferd G. Packer, Cannibal! Victim? (Frederick, Colo.: Platte 'n Press, 1980).

What Bailey needed was evidence to support Packer's story. For this he turned to the primary sources, scrutinizing every shred of evidence that had survived the passage of time. Most were preserved in the Hinsdale County Museum. The key to clearing Packer's name for good, Bailey felt, lay in a rusty, dented 1862 Colt pistol alleged to be the weapon that Packer used to shoot Bell. It had been found by a rancher decades ago in the massacre area. The five-shooter still had three bullets in the chamber. (Packer always said he shot Bell twice in the midriff.)

Bailey was convinced that Starrs got it wrong, believing that the mysterious hole in the skeleton's hipbone was caused not by gnawing animals but by a bullet. "I'm trying to get proof, on every level, to show Packer was attacked," Bailey said. "This evidence would put the last nail in the coffin."[16]

Over a period of time Bailey assembled his own team of experts, made up of chemists, soil and gunpowder experts, and an archaeologist, all with a mission to prove that Packer was telling the truth all along.

In February 2001 fragments of clothing and buttons recovered from beneath the bodies, together with soil samples from the grave, were sent for analysis at Mesa State College in Grand Junction, Colorado. The debris were combed to see if pieces of lead and unexploded black powder were present. Bailey explained that he was "looking for close fire gunshot residue."[17]

Professor Rick DuJay placed the samples borrowed from the Hinsdale County Museum on forty button-size pieces of carbon and studied them under an electron scanning microscope. DuJay was under no illusions about the enormity of the task facing him. In lay terms, scanning these samples was equivalent to searching every speck of dirt in an area of half a square mile. As he put it: "It's as if 127 years ago someone hit a baseball somewhere in the U.S. and now you're asked to find it."[18]

Miraculously, he did just that. After two days of intensive analysis, working at magnification powers of 1,000×, DuJay and his colleague, Professor Rex Cole, were able to visually identify a fragment of what appeared to be lead. Spectrographic analysis, which measures elements, confirmed that it was, indeed, pure lead.

Bailey echoed the team's delight. "The scientists . . . were astounded that it showed up this quickly."[19]

Next came the crucial test: How would this lead sample compare with the unfired bullets in the Colt revolver believed to be Packer's?

The spikes on the spectrograph matched exactly.

Ever cautious, the team wondered if the lead used in cartridges was spectrographically generic, so they took bullets from many eras, including the Civil War right up to the modern day and subjected them to an array of tests. Whereas all the other samples had anomalies such as tin or sulfur, the lead under the body and the lead in the gun were identical. Bailey was ecstatic. "Finding this chunk of lead is a great start. It matches lead in the bullets in the chamber of Packer's gun."[20]

What these tests cannot do is provide unequivocal proof that the lead actually came from a bullet. At the time of writing, research in this area is continuing.

Whether it will ever be possible to state indisputably that Alfred Packer really was a murderer is a question that will probably never be answered for sure, but in one respect he is unique. He remains the only cannibal to have his bust displayed in a state capitol building—so far as we know.

Chapter 4
Donald Merrett (1926)
Freed by Forsenics to Kill Again

At the dawn of the twentieth century, forensic science finally came of age. No longer was it the preserve of academics, to be detailed in dusty medical journals; it entered the mainstream. There were many reasons for this. Science was new, science was exciting, and as an increasingly educated public devoured what would nowadays seem staggeringly exhaustive newspaper accounts of criminal trials, the advances in laboratory techniques that were featured in so many of these trials must have seemed well-nigh miraculous. Unsurprisingly, the architects of this progress often achieved a near-mythic status, and nobody—certainly no one in Britain—benefited more from this newfound public awe than a lofty, rather austere doctor named Bernard Spilsbury.

Spilsbury was the first and greatest forensic superhero. At this distance of time it is difficult to comprehend just how famous he was, but in his heyday this sober-sided pathologist with his wire-framed spectacles and his stern, unyielding expression enjoyed a recognizability factor that put him on a par with politicians, movie stars, and top sports figures. When newsboys stood on street corners and bellowed at the top of their lungs, "Orrible murder—Spilsbury called in," they did so in the certain knowledge that that day's edition was a guaranteed sellout.

Success came early to the young physician. After gaining his medical qualification, Spilsbury went straight into forensic pathology in 1905, and within five years had reached the pinnacle of his profession, becoming senior pathologist at the British Home Office. He achieved this not only by being an outstanding diagnostician, meticulous and thoughtful, but even more important, at least as far as the authorities were concerned, through his magnificent bearing on the witness stand. Assured, authoritative, implacable, and above all absolutely convinced of his own infallibility, Spilsbury was a terrifying opponent for defense counsel and defendant alike. Juries literally hung on his every word. His testimony, delivered with an air of Jesuitical certainty, doomed scores of killers to the gallows, and we can only speculate on how Spilsbury must have felt, staring across the courtroom at some wretch, knowing full well that when next they met, a few weeks hence, the killer would be freshly hanged and the witness would be poised with his scalpel.*

Fleet Street worshiped him—he was effortless, memorable copy—and it was largely through their hyperbolic efforts that, in 1922, an appreciatory nation bestowed a knighthood upon the great man. In short order the cry of "Call Sir Bernard Spilsbury," echoing along oak-paneled corridors, passed into courtroom legend, the inevitable precursor to a sharpening of pencils in the press gallery and a sinking of spirits in the dock. He bestrode the legal arena like a colossus, in a way that no one else has before or since.

Being a Home Office pathologist meant that Spilsbury was more or less obliged to restrict his court appearances to the prosecution side, but late in 1926 a communication from Scotland made him break the habit of a lifetime. There was a tricky case in Edinburgh, and the defense wondered if Spilsbury might provide some assistance. After studying the paperwork, he agreed. Because Scotland maintained its own legal and judicial system, entirely separate from that of England, Spilsbury could act as an independent witness without any conflict of interest, and as the file notes revealed, this was a particularly unusual case.

At about nine-thirty on the morning of March 17, 1926, Henrietta (Rita) Sutherland was in the kitchen at 31 Buckingham Terrace, Edinburgh, where she worked as a housekeeper, clearing away the last of the breakfast dishes when she was startled by the sound of a gunshot. What happened next is unclear. In her first police statement Henrietta said that she ran into the living room just in time to see her employer, Mrs. Bertha Merrett, who had been writing a letter at a desk, fall to the ground. A pistol lay on the bureau behind her. According to a later statement, to which she

---

*Spilsbury regularly performed autopsies on criminals executed in the three London prisons of Wandsworth, Pentonville, and Holloway.

swore in court, Henrietta heard the shot, followed by a scream and a thud, and then Donald Merrett, age eighteen, ran into the kitchen, shouting, "Rita, my mother's shot herself . . . I used up her money and she was worried about it."[1]

In this version Henrietta said she had followed Donald into the living room and had seen Mrs. Merrett, plump, middle-aged, and well-to-do, sprawled on the floor between the oval table and the bureau, bleeding from a head wound but obviously still alive.

Astonishingly, no one bothered to phone a doctor. Instead, Donald called the police. When, within a few minutes, Constables Thomas Middlemiss and David Izatt arrived, the already confused circumstances of the shooting degenerated into a Feydeau farce. Neither officer appears to have been experienced in violent crime, and neither seems to have had the first clue about crime scene preservation. When asked later to reconstruct the position of the gravely wounded woman when they found her, each recalled something different. Their statements over the whereabouts of the gun were just as contradictory. One version had the stubby .25-caliber Spanish-made automatic lying on the desk, while another placed it on the carpet. Middlemiss remembered picking it up from somewhere; he just wasn't sure where. Needless to say, any possible fingerprint evidence was hopelessly compromised from the outset.

In the midst of all this Keystone Kops nonsense, an ambulance at last arrived. While Mrs. Merrett, unconscious and bleeding, was being loaded onto a stretcher and hurried off to the hospital, Middlemiss applied himself to tactful, almost obsequious questioning of the two witnesses. Violent crimes in this fashionable district of Edinburgh were scarcer than January heat waves, and the nervous constable had no wish to tread on any upper-middle-class toes.

He didn't get much change from Donald Merrett. Tall and heavily built with a brooding manner that many women found fascinating, the youth exuded an air of worldliness far beyond his tender years, and he made it clear that he regarded any questioning as an irritating intrusion. He explained that he had been in the living room, reading a book, when his mother had suddenly produced a pistol and shot herself.

When investigating detectives arrived and heard this, there was a collective scowl. Attempted suicide was a punishable offense, and they were anxious to know what might have precipitated such a drastic step.

"Money worries,"[2] said Donald blithely, producing two letters his mother had received from the Clydesdale Bank, detailing shortages in her account.

Whereupon the detectives closed their notebooks: case over. Every bit as incompetent as their subordinates, they, too, failed to process the crime scene, preferring to accept Donald's story at face value. As for Mrs. Merrett, she was whisked off to the infamous Ward Three at Edinburgh Royal Infirmary, which, with its barred windows and locked doors, was the repository for all attempted suicides, pending possible criminal charges.

Later that evening, Donald phoned the hospital and inquired if his mother was still alive. Informed that she was, the fun-loving young student and his girlfriend then took in a movie before checking into a hotel for the night.

A preliminary examination of Mrs. Merrett had revealed a bullet lodged at the base of the skull. An operation seemed out of the question, and yet, the next day, not only did she recover consciousness, but also she was quite lucid, if somewhat baffled as to how she had arrived at the infirmary. Since hospital rules forbade staff to discuss cases with suicides, she was told, "You had a little fall."[3]

This brought another puzzled frown from the patient. She had no memory of falling, only that "something like a pistol had gone off in her head."[4]

When someone asked, "Was there not a pistol there?" she seemed flabbergasted. "No, was there?"[5] She had some fuzzy recollection of Donald standing close by, and her saying, "Go away, Donald, and don't annoy me."[6] Then she heard the bang.

Bertha Merrett's bafflement carried over into a conversation with her sister, Eliza Penn. "Did Donald not do it? He's such a naughty boy."[7]

Despite her desperate plight, Mrs. Merrett's only concern was for her son, and she implored her sister to look after him. Eliza promised that she and her husband would. On the morning of April 1, Bertha Merrett died. Although never actually charged with attempted suicide, the harsh strictures of the age meant that she was denied the full rites of a church burial service. Her estate was left in trust to Donald until he was age twenty-one, allowing him to relax, secure in the knowledge that in a few years he would inherit a fortune of approximately £50,000 ($200,000), a vast sum for the period. In the meantime, to ease his grief, the bereaved son took off on an extended yachting trip.

Back in Edinburgh, the Penns were a worried couple. They had been doing some detective work of their own, and in the Merretts' living room, next to a wall, they found a spent cartridge case that the police had missed. Upstairs, in Donald's bedroom, was a box containing another thirty-eight cartridges of the same type. When Donald returned from his vacation he

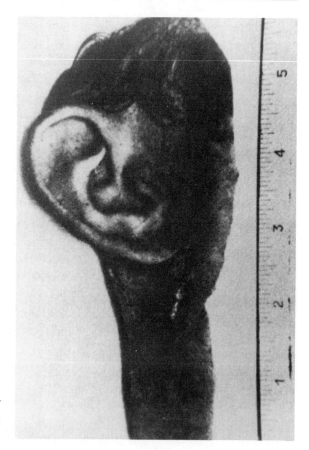

Sir Bernard Spilsbury's blunder over this ear allowed a killer to escape.

admitted buying the cartridges and the Spanish pistol just a few days before the tragedy so that he might shoot a few rabbits—hardly pests on the streets of Edinburgh—only for his mother to confiscate the weapon.

Eliza's qualms turned to outright suspicion when she learned that shortly before his mother's death, Donald had been living a double life. Instead of attending the local university, he had been seen roaring around Edinburgh on an expensive motorcycle, visiting dance halls and bedding a string of girls. Somehow he'd managed to fund all this hedonism on a weekly allowance of just ten shillings ($2).

It didn't take long for the source of Donald's newfound affluence to emerge. The trustee of Mrs. Merrett's estate, in struggling to untangle her chaotic finances, discovered incontrovertible evidence that Donald had been forging checks on his mother's account, to the tune of £457 (approximately $1,800). All through the summer Eliza Penn fretted. Clearly, Donald had been defrauding his mother. Had he murdered her also?

Doubts Begin to Emerge

Someone else mulling over this very same conundrum was Harvey Littlejohn, professor of forensic medicine at the University of Edinburgh, one of Scotland's finest pathologists. It was he who had performed the autopsy on Bertha Merrett. At the time Littlejohn had noted "a perforating wound"[8] behind the victim's right ear, made by a small-caliber bullet that had pierced the lobe and on entering the head had run along the base of the skull without damaging the brain. Death had occurred from infection of the wound, which had produced meningitis. Littlejohn had concluded: "There was nothing to indicate the distance at which the discharge of the weapon took place, whether from a few inches or at greater distance. So far as the position of the wound is concerned, the case is consistent with suicide."[9]

Now Littlejohn wasn't so sure. Rumors about the wastrel son had also reached his ears, as had news of the fraud, and he began to fear that advancing years—he was sixty-five—and increasingly ill health had marred his judgment. In Littlejohn's defense it should be noted that the police, eager to dispose of this case as expeditiously as possible, had impressed upon him their belief that Mrs. Merrett had committed suicide. Given that hindsight is always 20/20, it is easy now to cluck disapprovingly about Littlejohn's lack of thoroughness, but like many medical examiners, he was burdened by a high caseload, and in the absence of anything obvious to suggest otherwise, he found undeniable signs of the suicide that the police obviously wanted.

But had he blundered? In the unforgiving world of academia, where reputation is king and the knives are razor-sharp, it is a hardy soul who will admit to a doubt, let alone a mistake, and it is to Littlejohn's eternal credit that he bravely set aside personal considerations and decided to reappraise the evidence—this time as possible murder.

He spoke first with the two doctors who had attended Mrs. Merrett in the hospital almost continuously from her admittance until her death, and learned that both men had always doubted the police version of events. Working the suicide wing of the Royal Infirmary had made them familiar with gunshot wounds, and neither had noticed a smell of burning in the wound, or any powder blackening of the kind usually present in suicidal gunshot.

Littlejohn was aghast. All his professional life he had preached a mantra of skepticism, constantly warning students against the dangers of preconception; now he'd fallen headlong into that very same trap. Too ready to accept the police version of events, he had neglected to ask the attending doctors if there was any external evidence of a shot fired at close range. Such a fundamental lapse in judgment was unforgivable as far as Littlejohn was concerned, and it plunged him into a deep and dark depression. Riddled with chronic asthma and bronchitis, he began to experience pangs of worthlessness, made worse by fears that his ineptitude might allow a cold-blooded killer to walk free.

As he thrashed in gloomy despair, suddenly a chink of light appeared.

Few people in the world knew more about gunshot wounds than forty-three-year-old Professor Sydney Smith. In 1914 this brilliant pathologist had left Edinburgh, where he had studied under Littlejohn, to take up a post as principal medico-legal expert to the Egyptian government. There, the volatile social climate provided the hundreds of murders that would form the raw material for his textbook *Forensic Medicine and Toxicology*,* which caused a great stir with its revolutionary insistence that ballistics should be treated as a branch of forensic medicine. At the heart of Smith's ballistics philosophy was the principle of replication: when conducting experiments it was imperative to reproduce as closely as possible the parameters of the original shots.

This led to some groundbreaking discoveries. For instance, he found that to obtain the most accurate results from a revolver, it was necessary for the cartridge to be fired from the same chamber of the drum as had been used for the murder or suicide bullet. Another significant discovery was that powder blackening around a bullet hole would differ according to the age of the powder.

Coincidentally, Smith was visiting Edinburgh that summer, and he decided to call on his old friend and mentor. He was mortified by what he found. Littlejohn, normally so self-assured and ebullient, was a shadow of his former self, mired in self-doubt over the Merrett case. Like a man in a confessional, Littlejohn unburdened his soul. After absorbing the details, Smith agreed that murder was the likeliest cause of death. And he had a suggestion. "Why don't you make some experiments with the weapon that killed her, find out if a discharge close to the skin would cause powder-marks?"[10]

Shaken out of his torpor by Smith's encouragement, Littlejohn threw himself into a string of test firings with the Spanish automatic, using the same "Eley's Smokeless Cartridges" that had been recovered from Donald Merrett's room. By firing at moistened white cards, he found that at close range—up to three inches away—intense blackening and tattooing

---

*London: Churchill, 1925.

occurred, at six inches there were traces of blackening, and at nine inches none at all. This was highly significant. After all, what were the chances of a suicide holding a gun at least nine inches from the head? Although labeled "smokeless", the cartridges did, in fact, leave traces that could easily be seen by the naked eye. These findings led Littlejohn to conclude that in swabbing Mrs. Merrett's wound with warm water, hospital staff must have washed off the blackening.

However, when Littlejohn washed the test cards carefully, he found that some of the blackening and some of the impregnated powder could not be removed, except by repeated washings. It was inarguable: The pistol fired at close quarters left stubborn traces that no experienced doctor could have missed.

Littlejohn contacted the Procurator Fiscal of Midlothian County, Scotland, William Horn, admitted that his first report had been superficial, and declared that the possibility of suicide could be excluded. Delighted though he was by this development, Horn was caught in a quandary. Any defense counsel worth his salt would seize upon this volte-face like manna from heaven and offer it to the jury as irrefutable proof of prosecutorial weakness. Aware of just how unsympathetic juries can be to changes of heart on the witness stand, no matter how well argued or well intentioned, Horn needed more. For this reason he contacted Professor John Glaister, the distinguished head of forensic medicine at Glasgow University, and asked him to check Littlejohn's findings. After considerable discussion, Glaister agreed.

Working in tandem, Glaister and Littlejohn repeated the latter's experiments with the white cards, as well as firing bullets into human skin, using a leg that had been amputated at a local clinic. The results matched Littlejohn's earlier test results in every detail. Even after two months soaking in water, the skin sample still showed traces of blackening, whereas Mrs. Merrett's right ear—which along with a section of her scalp and neck skin had been excised from her corpse as evidence—showed none. Littlejohn contacted Horn and said, with Glaister's full agreement, that accident was inconceivable, suicide was in the highest degree improbable, and that all the circumstances pointed to the conclusion that the weapon had been fired by another party.

This was sufficient for Horn. Ongoing police inquiries into Donald Merrett's activities had exposed him as a serial fraudster who had been bilking his mother for months. Had his dishonesty been discovered, there was a good chance he might well have been cut off without a penny. Motive enough for murder? Horn thought so. In November he issued a formal petition to the Crown, charging Merrett with the murder of Bertha Merrett and with "uttering" forged checks. Arrested in the south of England, the prodigal son was escorted back to Scotland to stand trial.

The prosecution was sure it had the evidence to hang Donald Merrett. Littlejohn's tests looked ironclad. Then came stunning news: Merrett's defense team had engaged the services of none other than Sir Bernard Spilsbury.

Littlejohn reeled. For a man in ailing health and suffering an almost terminal attack of professional jitters, this was a real body blow. England's most formidable medico-legal expert had thrown down the gauntlet, prepared to cross swords with his Scottish counterpart in what promised to be a forensic battle royal.

Just the mere fact that Spilsbury—*the* great hanging witness—had agreed to appear for the defense added enormous weight to Merrett's cause. People who knew nothing about the case, and even less about ballistics, read their newspapers and shrugged their shoulders. "Well, if Sir Bernard Spilsbury says so, then the laddie must be innocent" seemed to be the popular view.

Littlejohn, meanwhile, checked and rechecked his results and could find no flaw. Let Spilsbury do his best—or his worst. He was ready for him.

Women Flock to Trial

The trial began on February 1, 1927. Judging from the number of women in the public gallery, Donald Merrett's air of coarse magnetism was still intact. They were drawn by a muscular teenage defendant without any hint of immaturity in his bearing, confident and erect. At times during the first couple of days his attention appeared to wander occasionally as neither side made any great headway, just engaging in preliminary sparring. But like everyone else, his concentration sharpened to a needle point when, on the third day, Professor Harvey Littlejohn was called to the stand.

The strain of the preceding months was painfully evident on Littlejohn's face as he entered the witness box. He looked ghastly. But he was a hardened courtroom veteran, and he soon settled into a good rhythm, giving his testimony in a clear, strong voice, describing the experiments that had led him to his conclusions. Everything, he said, hinged on the absence of smoke blackening around the bullet wound in question.

Nobody paid closer heed to this testimony than Merrett's defense counsel, Craigie Aitchison, one of Scotland's finest advocates, and when he rose to cross-examine the witness, it was clear that he was meticulously prepared.

Was it not true, he asked, that Littlejohn had contributed a foreword to Sydney Smith's *Forensic Medicine and Toxicology*? Yes, it was. Therefore, said Aitchison, it was safe to assume that Littlejohn considered Smith's book to be accurate and reliable? Again the witness agreed. In that case, Aitchison said, he would read a paragraph from this book, in which Smith, an internationally recognized expert on gunshot wounds, had written: "Although automatic weapons produce wounds identical with wounds from revolvers, it should be remembered that automatic ammunition is always charged with smokeless powder, and absence of blackening and burning in close discharges is relatively common."[11]

Littlejohn, far too savvy to fall into this carefully laid trap, politely suggested that counsel might care to continue reading the paragraph, in which Smith wrote that no hard-and-fast rules could be stated, that in every case experiments had to be conducted with the given weapon and given ammunition. His own experiments had shown him that the allegedly smokeless powder in Merrett's cartridges *did* leave distinct traces.

Aitchison blustered and quickly began quoting from other forensic textbooks, only to be pulled up short when Littlejohn politely pointed out that those volumes were now obsolete.

Aware that he was losing control of the exchanges, Aitchison launched a full-scale counterattack. Even if smokeless powder could leave traces, he snapped, it was well documented that these traces could be washed off. As support for this argument he brandished Littlejohn's own textbook, *Forensic Medicine*,* written just two years earlier. Surely the witness was prepared to stand by what he himself had written?

Naturally, said Littlejohn. Then Aitchison turned to page 120 of the book and photographs showing a bullet entrance wound in a suicide before and after cleaning. Alongside, Littlejohn had written: "With a sponge, the blackening of the smoke and any blood can be wiped off."[12] Did Littlejohn stand by this statement?

The witness said yes—and immediately suggested that, once again, Aitchison read the paragraph in its entirety, especially the conclusion that the powder blackening could be washed off, but not the particles of powder with which the wound was impregnated. He added to Aitchison's misery by further advising him to examine the second photograph on page 120 with a magnifying glass, when the particles of powder would be plainly visible.

It was a bravura performance from Littlejohn. He spared no one—least of all himself—in his determination to rectify a past error and emerged

---

*London: Churchill, 1925.

from the battle as that rarest of legal animals, an expert witness prepared to publicly admit to fallibility. His evidence had been strong and convincing, and when Glaister, in his testimony, forcibly reiterated every one of Littlejohn's conclusions, it should have been game, set, and match for the prosecution. But Aitchison was ready to serve a thunderous ace.

"Call Sir Bernard Spilsbury," he told the clerk of the court.

Massively imposing and immaculately attired, with the inevitable carnation in his buttonhole, Spilsbury took his place in a witness box that suddenly seemed too small to accommodate the great man. Even Aitchison fell victim to his awesome presence, causing titters of amusement when he inadvertently addressed the witness as "St. Bernard."

Spilsbury's domination began from the very first syllable. Unprompted and with scarcely any interruption, he launched into an account of having heard of the ballistics experiments in Edinburgh, and deciding to make trials of his own. He had been joined in this endeavor by an old colleague, Robert Churchill, a London gunsmith of international repute. "The experiments were made with Mr. Churchill, using an automatic pistol as nearly as possible like the one which was used in this case. We had a description of the pistol and we picked one having exactly the same length of barrel and the same bore. We also selected cartridges which corresponded as closely as possible to the description we received of those which were used in this case."[13]

A spellbound hush filled the courtroom as Spilsbury continued what amounted to a lecture. It all sounded wondrously impressive until one realized that with every word the witness was undermining the value of his own experiments. As Smith and others had already discovered, a different gun and different ammunition made for different results, so it came as no surprise when Spilsbury revealed that his findings had varied greatly from Littlejohn's. Before getting to specifics, he gave the court the benefit of some general observations. The site of the wound behind the ear was, he said, entirely consistent with any suicide, "let alone a woman,"[14] who, he thought, would involuntarily flinch away from the coming explosion at the moment of pulling the trigger.

He went on to say, "I have applied my mind to the question of the possibility of drawing any certain conclusions as to whether the wound which resulted in Mrs. Merrett's death was homicidal, suicidal, or accidental."[15]

At this point Spilsbury resorted to one of his favorite courtroom devices, a reference to his "own experience." Plucking examples from the past to bolster the testimony of today was a ploy he fell back on repeatedly throughout his career, knowing full well that no counsel would dare challenge his integrity by demanding documentary corroboration. Spilsbury guarded his reputation jealously and was secretive to the point of paranoia. He never published his results, he never wrote any textbooks, he never shared his hard-won knowledge with students, and he never divulged any verifiable statistics. The details of the long career—more than twenty-five thousand autopsies—were recorded for his eyes only in the card index locked away in his Gower Street, London, office.

On this occasion, Spilsbury cited from his own records the case of a suicide who had also shot himself behind the right ear. This allowed him to conclude: "So far as my experience goes, there is nothing in the site of the wound in the case of Mrs. Merrett inconsistent with suicide."[16]

He next turned his attention to the automatic pistol. He pointed out that, although of cheap make, the weapon was in good working condition, light in weight, with a trigger pull of just six pounds. Its very short barrel, projecting just one and a quarter inches beyond the finger on the trigger, made it easy to hold it in a position against the side of the head. Owing to the light weight there would be no strain on the hand or arm even if the pistol was held two inches away from the head.

Now Spilsbury had to explain the absence of blackening in the wound. He stated that no definite inference could be drawn concerning the distance of the pistol from the head, because bleeding and swabbing of the wound would have washed off any powder marks. And he summarized the experiments he had conducted with Churchill by exhibiting cards that showed part of the blackening wiped off by a damp cloth. At two inches there was very little blackening. On human skin, it would be even less distinct, said Spilsbury, because the nature of the surface—its moisture or greasiness—made the removal of such markings much easier. After wiping the skin, a magnifying glass would be needed to detect any trace.

Another vital consideration was the quantity and type of propellant used in the bullets that Spilsbury tested. Just two grains of flake smokeless powder—a very small charge, indeed—went into each .25 cartridge, and flake, unlike granulated powder, would not readily penetrate the skin, but instead adhere to it, which meant that any blackening from the discharge would be superficial and easily removed by a damp cloth. Spilsbury was emphatic. Nothing in his London experiments had indicated that it was possible to deduce how far the gun had been from Mrs. Merrett's head when it was fired.

Spilsbury Slips Up

Thus far Spilsbury had been his usual, impeccable self. Now he proceeded to dig something of a hole for himself. He told the court that, far from relying on his London experiments alone, he had also carried out test firings in Edinburgh, with Merrett's own gun. Unfortunately, he had used cartridges bought in London, not the original ammunition. Also, he had fired only at paper, not skin, and made no attempts to remove the powder marks. No attentive listener could fail to have been startled when Spilsbury announced that the experiments in Edinburgh had resulted in much more blackening than those in London. In other words, even using the same ammunition in a different gun had produced radically different results. This was the clearest vindication yet for Professor Smith's still controversial theory that for ballistics test results to be as accurate as possible, the original weapon and the original ammunition should both be present.

Aitchison rushed to repair the damage. "As regards the ease with which blackening could be removed, was there any difference between the London experiments and the Edinburgh experiments?"[17]

"I should say not,"[18] Spilsbury replied in his most lordly fashion, adding that the experiments in Edinburgh had in no way altered his conclusions.

But the chief prosecutor, William Watson, had noted the flaw in Spilsbury's experiments and was champing at the bit when it came time to cross-examine. Did Spilsbury agree that comparison experiments should properly be made with the original gun and the original cartridges? "Of course,"[19] replied Spilsbury. Whereupon Watson pointed out that the experiments in London had not fulfilled this requirement. In view of this fact, he asked, did not Spilsbury agree that Littlejohn's experiments therefore had more validity than his own? Icily and with his customary composure, Spilsbury repudiated this suggestion.

As he neared the end of his testimony, a rare note of circumspection crept in as he reiterated his opinion that although suicide was one possible explanation of what had occurred, an accident could not be ruled out entirely. "From my own experience of accidental shooting one knows of extraordinary positions, sometimes resulting from the accidental discharge of weapons, and in such a position as this an accidental discharge, I think could never be entirely excluded. It might even be at a range greater than that which would produce local marks around the wound."[20] Significantly, the possibility of murder was just brushed to one side and barely mentioned.

In choosing not to testify, Donald Merrett may have irked the public gallery, but probably saved his life. Had he been exposed to the full rigors of cross-examination, it's unlikely that even Spilsbury's testimony could have saved him. As it was, Merrett chose to put his fate in Aitchison's hands, and the lawyer didn't disappoint. Without once impugning the status of such skilled homegrown talent as Littlejohn or Glaister, Aitchison emphasized that in the whole field of forensic science in Britain and in Europe as well, the name of Spilsbury was a byword for knowledge, integrity, and, above all, infallibility. The man who virtually invented the role of "expert witness" had spoken for the defendant, and, Aitchison argued forcibly, that should be enough for any jury.

This was a theme hammered home by the judge in his summing up. Referring to the welter of contradictory forensic testimony, Lord Alness made his own prejudices clear to the jury, stressing Spilsbury's conclusions, to which, he said, "I should imagine you would be disposed to attach the very greatest weight."[21] Then the jury of six women and nine men retired. An hour later they were back with that uniquely Scottish hybrid verdict "not proven" to the charge of murder, which roughly translates to, "We think the defendant did it, but the evidence wasn't quite there." On the charge of check forging, they were unanimous: guilty. For this crime Merrett received a one-year sentence.

When later told the verdict by Littlejohn, Professor Sydney Smith shook his head sagely and remarked, "That is not the last we'll hear of young Merrett."[22]

Sadly, Littlejohn had no opportunity to test the wisdom or otherwise of Smith's prophecy. The Merrett case had taken a terrible toll of his health, and was, according to Smith, "a contributory factor in his early death"[23] just six months later, in August 1927.*

Although Sydney Smith did not testify directly at this trial, it was his first contact with Spilsbury, and he didn't much care for what he saw. The cult of personality that had grown up around the legendary pathologist—the product of years of unalloyed praise and virtually no opposition—had inflated his ego to the size of a Zeppelin, with dangerous consequences for the interests of justice.

This syndrome reached its apotheosis in 1944. Coming toward the end of what had been a long and strenuous career, Spilsbury was a tired man

---

*Upon Littlejohn's death, Sydney Smith took the vacant chair of forensic medicine at Edinburgh University.

when he agreed to once again testify for the defense at a capital murder trial. Harold Loughans, a career burglar with a mutilated hand, had been charged with strangling a pub landlady during a botched robbery, but retracted his confession in court, claiming it had been a police invention. On the basis of one flabby handshake with the defendant, Spilsbury testified, "I do not believe he could strangle anyone with that hand."[24]

When an incredulous prosecuting counsel asked if it was possible that Spilsbury had been hoodwinked by a killer desperate to escape the noose, he replied disdainfully, "No, I don't think so."[25] And that in a nutshell was Sir Bernard Spilsbury, utterly incapable of admitting that he might be wrong. So what if the police and the judiciary realized that Spilsbury's powers were declining? His hold over the jury was impregnable, and, sure enough, Loughans was acquitted. Another two decades would pass before the truth of this case became known. In 1963, immune to the threat of further prosecution, Loughans finally came clean, admitting in a newspaper interview that he had, indeed, strangled the pub landlady.

Spilsbury's own death on December 17, 1947, by his own hand, spared him any embarrassment he might have felt. It also prevented him from learning what became of the strapping eighteen-year-old he had defended so resolutely in an Edinburgh courtroom twenty years earlier.

Life certainly hadn't been dull for the buccaneering Donald Merrett: a quick marriage; another jail term, for jewelry fraud; triumph in 1929 when he finally got his hands on that trust fund; high living through the 1930s; and action on torpedo boats during World War II, which saw him captured by the Italians, only to escape and fight again. After the war his sense of adventure and his larcenous nature steered him into the life of a black marketeer, smuggling cigarettes and whiskey in the Mediterranean, but the prison sentences kept cropping up and so did the debts. By the early 1950s the by now grossly overweight desperado was also in perilous financial shape. He'd squandered every last penny of his inheritance, and the debts were piling up. Penniless in West Germany, he kept hankering after the £8,400 (approximately $33,000) he had so rashly settled on his young wife in 1929, money that would revert to himself in the event of her death.

That sad occasion occurred on February 11, 1954, when Merrett, now calling himself Ronald Chesney, secretly returned to England from West Germany on a false passport, plied his wife with enough gin to sink a battleship, then drowned her in a bathtub at her West London home. He might have escaped justice once again, had it not been for the fact that as he was sneaking down the stairs his mother-in-law suddenly appeared, clutching a tray of coffee. To silence her screams, Merrett bludgeoned her about the head with the coffeepot, then fled back to West Germany.

Unfortunately for Merrett, the screams had attracted neighbors, who spotted his hulking frame as he lumbered off down the road. Hearing that Scotland Yard were pursuing him for double murder, and resigned to certain execution if caught, on February 16 Donald Merrett, alias Ronald Chesney, went to a park in Cologne and shot himself through the head. Bloodstains and hair found on his suit linked him conclusively to the murders in London, where an inquest decided that he had unlawfully killed his wife and mother-in-law.

So finally a line was drawn under the name of Donald Merrett, twenty-seven years too late. Spilsbury's obduracy not only cost the lives of two more women, it also put a deep stain on his Olympian reputation and highlighted the dangers of forensic idolatry. His passing triggered an undeniable change in attitude in the courtrooms of Britain. Gone were the days of forelock-tugging sycophancy that Spilsbury had regarded as his by right, supplanted by more rigorous examination and a need for documentary evidence. Perhaps even more significantly, it strengthened the realization that who among us, no matter how brilliant, how well qualified, how experienced, is entirely immune to an occasional error in judgment?

Chapter 5

William Lancaster (1932)

A Bullet in the Night

Junk science has always been with us. Right from the days when medieval con men bragged of having found the "philosopher's stone" that would turn base metal into gold, hucksters have, under the guise of scientific expert, been passing opinion off as fact, or else bending test results to their own purpose. It has been no different in the field of forensic science. Whether from a misguided sense of loyalty, hubris, or just plain greed, a disturbing number of expert witnesses have been guilty of economizing on the truth when testifying in court.

In recent years the infamous Fred Zain wreaked havoc wherever he worked. As chief serologist for the West Virginia crime laboratory, Zain's testimony sent hundreds of defendants to prison, and he did likewise in Texas, when he moved to San Antonio. Misgivings about his integrity crescendoed in the early 1990s and led to an investigation by the West Virginia Supreme Court, which found that serious doubt existed in 134 cases where Zain had testified. Eventually nine men were freed. Zain was subsequently fired from his position in Texas and found himself facing criminal charges.

Also in Texas, pathologist Ralph Erdmann, who worked as a contract medical examiner in forty counties, was caught having faked more than a hundred autopsies on unexamined bodies and having falsified dozens of toxicology and blood reports. Dozens of other autopsies were botched. In one memorable case he even lost a head. For all this crookedness and incompetence, he received just a ten-year probation order and some community service.

Elsewhere, a 1992 rape trial conviction was thrown out after a Chicago police lab analyst, Pamela Fish, was found to have excluded exculpatory serological results in testimony. (In 1999, the defendant was exonerated by DNA tests.)

Another controversial figure is Joyce Gilchrist, a supervisor in the Oklahoma City police laboratory since 1980, someone who once boasted, "I seemed to be able to do things with evidence that nobody else was able to do."[1] She was as good as her word. Her testimony sent twenty-three men to death row, eleven of whom have been executed. Yet as far back as 1987, colleagues were expressing concerns about her competence, and she was given a warning by one professional association and expelled by another. Despite this, she was promoted and kept testifying in capital cases. The full extent of her ineptitude didn't became apparent until 1998, when Robert Lee Miller, after spending ten years on death row for a murder-rape, walked free when new DNA evidence proved that Gilchrist had misidentified a semen sample (the real killer was subsequently convicted). When asked to peer-review Gilchrist's work, an FBI investigation found that she gave misleading testimony in five cases and that the matches she

PRINCIPALS IN MIAMI'S SENSATIONAL LOVE TRIANGLE

HADEN CLARKE

CAPTAIN LANCASTER

MRS. JESSIE KEITH-MILLER

Bill Lancaster's sensational trial garnered international headlines.

made fell "fall far below the acceptable limits of the science of hair comparisons." They also criticized her for "errors in identification."[2] At the time of writing, Ms. Gilchrist is suspended.

In Britain, too, problems exist. Multiple convictions involving alleged IRA terrorists had to be overturned after gross irregularities were discovered in the test procedures of one particular government scientist, Dr. Frank Skuse. And the whole sorry farrago goes on.

Whereas all these regrettable examples hinge upon bias and flawed methodology, back in the dark ages of forensic science—the late nineteenth and early twentieth centuries—the core problem with expert testimony was money. Dump enough cash on the table and you could be sure to rustle up some shyster ready to spout a stream of psuedoscientific mumbo jumbo designed to present your case in the best possible light and pull the wool over the jury's eyes. From out of what was a pretty crowded field, one man emerged as the undisputed master of evidence manipulation. Perhaps his greatest "success" came in a sensational South Florida murder trial.

On May 21, 1927, an exhausted young American airman stepped down from his fragile monoplane at Le Bourget Airfield in Paris and into aviation history as the first person to fly the Atlantic nonstop and solo. Within twenty-four hours Charles Lindbergh was the most famous man on earth, feted and lionized in every corner of the globe. And the world had a brand-new spectator sport: aviation. Many of the pilots became household names, as famous as movie stars, and there were riches in abundance for those brave enough to tackle the skies in search of new aeronautical records.

In England one pilot bitten by the Lindbergh bug was an ex-Royal Air Force officer, Captain William Lancaster. Since leaving the service in 1926, Lancaster had struggled in civilian life, and it showed. The thinning hair and cadaverous features made a mockery of his real age—twenty-nine—giving him the appearance of someone much older, while his bank account was equally worn out, exhausted by the effort of supporting a wife and two daughters on erratic, often nonexistent earnings. Not that it troubled Bill. Blessed with bottomless reserves of optimism, he shrugged off each setback like it never happened, all the while keeping his gaze firmly fixed on his big dream—being the first person to fly a light plane from Britain to Australia.

There was just one problem: record attempts were fantastically expensive, and Bill couldn't find financial backers anywhere. And then, in June 1927, at a London cocktail party, amid the chinking glasses and the gaiety, he found himself buttonholed by a persistent young Australian woman, Jessie Keith-Miller.

Called "Chubbie" by her friends, on account of the fact that she scarcely weighed a hundred pounds, the diminutive twenty-five-year-old was a pocket dynamo of drive and vivacity. She was separated from her husband and ravenously hungry for excitement, and she pledged to help Bill raise funds for his flight and find a plane. There was just one proviso—she wanted to go along.

Bill immediately agreed. Female fliers—or aviatrixes, as the press termed them—were rare jewels, indeed: Chubbie's participation would be bound to generate gales of publicity. And so it proved. Hordes of reporters and photographers converged on Croydon Aerodrome, just south of London, on October 14, 1927, to see the departing couple as they posed confidently in leather jackets and flying goggles beside their single-engined Avro Avian named *Red Rose*. Also there to see them off was Bill's long-suffering wife and one of his daughters. Kiki Lancaster put a brave face on what was undoubtedly a gut-wrenching moment as she watched her husband fly off into the clouds. It would be years before she saw him again.

Under Bill's tutelage, Chubbie quickly learned to fly the tricky little biplane, down across Europe, across to Africa, into Asia. With a top speed of just eighty miles an hour, *Red Rose* taxed both endurance and nerves to the limit. Chubbie and Bill rose to every challenge. They skimmed just feet above basking sharks in the Arabian Gulf as fuel ran low; they fixed the cantankerous engine when it broke down; they battled blinding tropical rainstorms; they even fought snakes in the cockpit; and one night, alone beneath the Persian stars, they became lovers.

Then nemesis played its hand.

On January 9, 1928, more than eighty-five hundred miles into their trip, they crashed on the island of Sumatra. They were lucky to escape with their lives, but the plane was badly damaged, and as they kicked their heels waiting for spare parts to arrive, they had to endure the frustration of suffering as a young Australian, "Hustling" Bert Hinkler, overtook them and went on to claim the record that Bill had craved.

Although forced to make do with second best, it scarcely mattered. When the couple finally touched down in Darwin on March 19, 1928, Chubbie was lauded by her fellow Australians as a heroine. No other woman had flown so far in a light plane. Not even Amelia Earhart, who later that year would become the first female passenger to fly the Atlantic, could rival Chubbie's achievement. But whereas Earhart had the matchless promotional skills of her publisher husband, George Putnam, to turn her triumphs into worldwide fame and a bulging bank balance, Chubbie had Bill; and as always, Bill couldn't scrape two cents together.

Not that he didn't try. In Australia he negotiated with a U.S. movie company to film their epic journey, but when the couple reached Los

Angeles the promised movie deal fell through, forcing them back into the skies. When Chubbie gained her full flying license in New Jersey, she became one of only thirty-four female pilots in the United States, and, briefly, her star soared. Twice, in her monoplane called *Alexander Bullet,* she set records for transcontinental flights across America. She took part in several air races for women only—"powder puff derbies"—until crashing on Andros Island in the Bahamas on a round-trip race from Pittsburgh to Havana. Chubbie's dramatic rescue story, in which Bill played a big part, earned them $1,500 from a newspaper, but that was the last of the easy money. With the Depression biting hard, America's appetite for air races faded fast.

On a personal level, Bill was undergoing a deep crisis. He yearned to marry Chubbie, only to be stymied by his wife's refusal to grant him a divorce. Chubbie, freshly divorced herself, also was edgy, but for a very different reason—the flame she'd carried for Bill Lancaster across four continents was beginning to flicker and wane.

Beset by very different demons, each found solace in the bottle. In late 1931, broke and embittered, they drifted down to South Florida, renting a large and airy house in Coral Gables. Their fortunes continued to spiral downward, just a couple of broken-down fliers, drowning their sorrows in Prohibition gin, surrounded by hibiscus and bougainvillea and memories.

Fateful Meeting

A rare interlude in this drunken odyssey came in February 1932, when rumors of a flying job took Bill out of town. While he was gone Chubbie chanced to meet Haden Clarke, a dashingly handsome writer with dark wavy hair, bushy eyebrows, a solid jaw, and a wife somewhere out West. At first Chubbie found Clarke's oily insouciance distasteful, but her curiosity was piqued when he suggested ghostwriting an account of her adventures. Once he'd outlined his qualifications—a master's from Columbia and plenty of newspaper and national magazine experience—Chubbie was hooked. So was Bill, when he returned. In his diary he wrote, "Met Haden Clarke, a writer. First impression of him very good."[3]

A deal was struck. No money would change hands, but any royalties would be split down the middle. To facilitate research and help Clarke—who was flat broke—it was agreed that he should move into the house.

The sleeping arrangements were clearly demarcated: Chubbie had her own room upstairs; Bill and Clarke shared a sleeping porch over the garage. Otherwise, nothing had changed: utilities went unpaid; the rent was overdue; Bill poached chickens to keep them fed; and when the electricity was finally shut off, Chubbie was reduced to barbecuing over an open fire in the yard. Through it all, the booze kept flowing. Haden turned out to be a real lush, and Chubbie, too, was guzzling gin by the gallon. Only Bill made any effort to find work.

He discussed plans with a local company, Latin-American Airways (LAA), about establishing a Miami–Mexico air route. Although deeply suspicious of the company's real motives—he smelled more than a whiff of drug smuggling—Bill agreed to fly to Mexico to test the market. On his last night in Miami, painfully aware of the manner in which alcohol set Chubbie's passions racing, Bill took Haden to one side and begged him to keep her off the bottle. Haden promised to do his best.

The next day, March 6, 1932, Bill took off for Mexico.

The expedition proved to be a calamity. As Bill had suspected, the name of the game was trafficking: not drugs, but illegal Chinese immigrants. Trapped with a bunch of hoodlums and stranded two thousand miles from the woman he loved, he still managed to send back every spare cent, once slipping a single dollar bill into an envelope. He even pawned the company's gun for five dollars and mailed this off to Chubbie. All the while his imagination, rubbed raw by his own miserable plight, conjured up all sorts of lurid scenarios at the house in Coral Gables, none of them good.

He kept in touch via a blizzard of letters and telegrams. At first Chubbie sounded just grumpy—"There is also not a drop of booze in the house and we can't afford to buy any"[4]—and she'd followed this with a scathing attack on Clarke—"He is without doubt the laziest and slowest writer I have ever seen"[5]—but gradually the tone of her replies began to shift. Later, she wrote, "Don't hurry back. . . . Having spent so much time and money, not to mention hardships, it seems an awful pity to give up now. . . . Haden has been a peach."[6]

Indeed, he had. One morning Chubbie had awakened with a throbbing hangover to find Clarke asleep across the foot of her bed. Although she had been too drunk to even remember what, if anything, had occurred, this marked the beginning of a torrid affair. Haden, impulsive and romantic, begged Chubbie to marry him, and with Bill seemingly a million miles away, she agreed.

Mischief-making acquaintances kept Bill apprised of every development. One night in Mexico a supposed friend showed Bill letters he had received from his wife. "Chubbie and Clarke came round tonight . . . all ginned up. I really think now that Clarke has gained Chubbie's affections, and Bill lost them . . . Don't tell Bill, but I believe she is well satisfied."[7] The last two words were underlined.

Bill copied the letters into his diary and underwent the torments of hell itself.

After weeks of near starvation, he scrounged a job flying a plane to a friend in St Louis. When he landed on April 15, a letter from Clarke greeted him. "I have communicated with my wife and have made arrangements for an immediate divorce. Chubbie and I plan to marry as soon as it is granted."[8] Chubbie, too, had written Bill to say that it was all over.

Before leaving St. Louis, Bill wired Miami. "Am no dog in manger but hold your horses, kids, until I arrive. Insist on being best man and best friend of you both for life . . . love, Bill."[9]

Then, using borrowed money, he spent $30 on a .38 Colt revolver—to replace the company gun he'd pawned earlier, he explained later. That night, April 19, when he landed in Nashville, he loaded the revolver, and it was beside him the following day as he took off on the final leg of his journey. At seven o'clock that evening his plane touched down at Viking Airport in Miami.

Chubbie and Clarke stood waiting by the runway.

The homecoming was stiff and painfully polite. After exchanging courtesies they drove to the house. To Bill's utter amazement, when he suggested a drink, both Chubbie and Haden declined. Only later would the significance of this refusal become apparent.

Inside the stifling house, tempers flared over dinner as both men laid claim to the woman they loved. Clarke sulked, while Bill soon made it clear that his telegram was a sham; he had no intention of surrendering without a fight. Chubbie, caught in the middle, acted as fretful umpire.

Among the acrimony and recriminations came the blurted bombshell, the reason why Haden was abstaining from alcohol—he had contracted a venereal disease. Shamefacedly, Chubbie admitted that she had gone on the wagon in sympathy, and that she'd banished Haden from her bedroom until he was cured.

Bill erupted. He implored Chubbie to reconsider, to ditch this disease-ridden slacker and return to him. Although Chubbie refused to abandon her new lover, she did accede to Bill's request that no marriage take place for at least a month. Her concession did little to cool the overheated atmosphere. Both men kept sniping at each other. Eventually Bill stormed from the house.

He returned around midnight, ostensibly to gather a few things before checking into a hotel, but Chubbie talked him into staying, and all three made preparations for bed. Bill, it was agreed, would share the sleeping porch with Haden.

At 12:45 A.M. Chubbie retired to her room, having locked the door in accordance with Haden's wishes. He didn't want Bill attempting any nocturnal diplomacy.

She read for a while. In the background the voices of the two men filtered through the thin walls, not bellicose or shouting, just talking, even laughing at one point. With a sigh of relief, she switched off the light and fell asleep.

Sometime after 2:00 A.M. she was awakened by a fearful pounding at the door. A voice shouted for her to get up. She ran to the door. It was Bill, ashen-faced.

"A terrible thing has happened," he gasped. "Haden has shot himself!"[10]

Chubbie's first reaction was incredulity. "Don't talk nonsense. There's no gun in the house."[11]

While Bill garbled a breathless account about the gun purchase in St. Louis, Chubbie rushed past him, into the sleeping porch where Clarke lay on his bed, groaning but unconscious, his face streaked in blood. She got a damp washcloth and began wiping the blood from Haden's forehead, shouting for Bill to call a doctor. Only then did she notice the gun, partly under Haden's body, near the waist. With two fingers, she pulled the barrel out about an inch, then thought better of it.

When Bill returned from the phone he pointed to a table at the foot of Haden's bed and two notes, both typed and signed in pencil. One read: "Chubbie, the economic situation is such I can't go through with it. Comfort mother in her sorrow. You have Bill. He is the whitest man I know. Haden."[12]

The second read: "Bill, I can't make the grade. Tell Chubbie of our talk. My advice is, never leave her again. H."[13] Once Chubbie had read the notes Bill urged her to destroy them, to conceal the scandal of their love triangle, but she refused.

At that moment an ambulance man arrived. As he examined Haden, Bill hovered close by, eager to know whether Haden would ever speak again. The driver said he doubted it.

The next person to arrive at the house was Ernest Huston, an ex-lawyer for LAA, who had come in response to an urgent call from Bill. When Bill again suggested destroying the notes, Huston cautioned him not to, at which point Bill, wringing his hands, paced over to the dying, comatose man and said, "I wish that Haden would talk, so he could tell us why he did it."[14]

Haden never did talk, and at eleven-twenty that same morning, he died at Jackson Memorial Hospital. The next day he was buried, without even the formality of an autopsy.

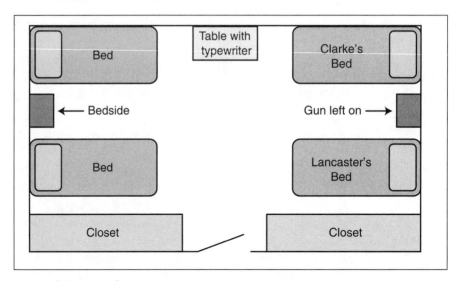

Layout of sleeping porch.

Given the bizarre circumstances of that night in Coral Gables, few were surprised when, just over three months later, William Lancaster sat in a Miami courtroom, facing charges of first-degree murder. Most considered the outcome a foregone conclusion. After all, even his own counsel, James Carson, when first approached to handle Lancaster's defense, had snorted, "He's as guilty as hell."[15]

It certainly appeared that way. Right from the outset everything had gone wrong for Bill Lancaster.

When taken back to the house that same night, he and Chubbie had attempted to conceal the suicide notes, but Detective Earl Hudson had already spoken to the ambulance driver and demanded to see them. While Hudson examined the notes, he asked if there was any reason why Clarke would shoot himself. Bill replied that he had been depressed, having just learned that he was suffering from a disease.

When Hudson inspected the sleeping porch he saw the .38 Colt still lying on Haden's bed. In an act of almost criminal stupidity, he reached forward with a handkerchief, picked up the gun, and slipped it into his pocket. Perhaps his attention had already been distracted by a pencil on the desk, apparently the same pencil used to sign the notes. It was smeared with blood.

Hudson frowned. How could Haden sign the notes, put down the pencil, go over to the bed, shoot himself, yet leave blood on the pencil?

Bill was ready with a quick answer: after finding Haden, he had seen the pencil on the floor and replaced it on the desk. Most likely his hands had been bloody from touching Haden's head.

Hudson wasn't convinced, and neither was the DA's office. Nor was their mood improved by news that fingerprints on the gun were too smudged to identify, almost certainly the result of Hudson's sloppiness. This only left the suicide notes. They were sent to document examiners, who were asked to compare them with examples of both Haden and Bill's typing, trying to establish who was the more likely to have written the notes.

Analyzing literary style for forensic purposes is a hugely subjective exercise, which probably explains their ambiguous findings, but on one point there was universal accord—the signatures were forged. Whoever wrote the name "Haden" had removed the pencil at least fifteen times in their painstaking attempt to duplicate the genuine signature.

Before the DA could act on this information, Bill tried a preemptive strike—he showed up at the Miami courthouse with an admission that, yes, he had forged the notes.

He told of lying in bed and being awakened by what sounded like a window banging shut. After fumbling for the light switch, he saw, just three feet away, Haden on his bed, blood oozing down his face. It took a moment or two for the horrific realization to sink into Bill's conscious-ness—his gun, left overnight on the nightstand between the two beds, now lay half buried beneath Haden's body.

After failing to elicit any response from Haden, Bill panicked, fearful that Chubbie would think he had shot his rival in a jealous rage. Hurriedly he typed the two notes, which he had then attempted to get Haden to sign. When this failed, he had signed the notes himself, then called Chubbie.

It didn't sound good—in fact, it stank—and on May 2 Bill was hauled off to jail on charges of murder.

Sex and Sin in Miami

His trial attracted worldwide publicity. It had everything: a juicy murder; high-profile principals; international interest; and lashings of illicit sex. Newspaper editors smacked their lips over rumors that Chubbie would go on the stand and blow the lid on the infamous Coral Gables "love nest." Hardly surprising, then, that when the proceedings got under way on August 2, 1932, Miami's courthouse was packed with reporters from as far afield as Australia and Britain.

Ever since the tragedy, Chubbie had been brutally torn. The more she had learned about Haden—he was just twenty-six, not the thirty-one he had claimed; his qualifications and writing credentials were bogus; he had not one but two wives simultaneously; and his venereal disease had been chronic, instead of the recent affliction she had been led to believe—the more she realized how thoroughly she had been duped. Even though her feelings toward Bill had not changed—that vital spark of love was forever extinguished—their shared experiences meant she felt bound to him by an unquenchable sense of loyalty. Dull he might be, but he was decent and dependable and, above all, truthful.

This would be James Carson's stance throughout the trial. As noted earlier, the lawyer entrusted with keeping Bill out of the electric chair had originally been convinced of his client's guilt; however, close personal contact had revised his opinion. Captain William Lancaster, the stalwart British flying officer, just didn't seem the type of fellow who carried out this type of crime. Dammit, well-bred gentlemen just don't go around shooting chaps, no matter how caddish and deserving the victim!

It was blatant snobbery, of course, and totally absurd, but Carson knew his juries: given the choice between a pukka British officer and some dope-smoking, alcoholic bigamist with the clap, he had a fair idea which way they would jump.

Before all that, however, there was a mighty forensic battle to fight.

The study of ballistics essentially falls into two distinct categories—the effect of the firearm on the projectile, hopefully leading to the identification of a particular weapon, and the effect of the projectile on the target. When the target happens to be human flesh, three more questions arise: Was the shot accidental, homicidal, or suicidal? Since accidental shooting was not an issue in this case, it left a straight choice.

Deciding whether a gunshot wound is homicidal or suicidal has always been a thorny topic. It is nowhere near as straightforward as one might imagine. The great enemy is blind assumption; after all, most people confronted by a body riddled by half a dozen bullets might instantly suspect murder, but autopsy records are packed with accounts of suicides peppering themselves with gunshots before succumbing to their wounds.

In this case a major problem had been the lack of an autopsy, an omission remedied in late May by a court order authorizing the exhumation of Clarke's body. The autopsy was overseen by a panel of doctors made up of both prosecution and defense experts. The massive radiating fractures on the left side of the skull—caused by the exiting bullet—led them to conclude unanimously that Clarke had been killed by a close-contact wound to the upper right temple.

Contrary to popular belief, not all close gunshot wounds leave those telltale powder burns on the skin surface so beloved of fiction writers. If the barrel is pressed hard against the skin, the escaping gases may go deep into the wound, searing along the bullet's track, as happened here. Sometimes a concentric impression is left around the wound, caused by the gases blowing the skin back against the muzzle of the gun.

Dr. Carleton Deederer, who had examined Haden at the hospital, had carefully noted the position of the entry wound. The bullet had entered the head halfway between the right eye and the right ear. The exit wound was above the left ear but higher, near the top of the head. This meant that the bullet had traveled diagonally, front to back, and upward through Haden's skull as he lay on the bed, before exiting and burying itself in his pillow.

This bullet track posed serious problems for the suicide theory. As Haden was shot while lying with his face toward Bill's bed, it would have meant that he took the Colt revolver from the nightstand, either held or balanced the heavy gun almost vertically above his head, and angled it back toward himself before somehow squeezing the trigger. Even setting aside the unlikelihood of such self-immolatory gymnastics, the majority of the panel doubted whether this course of action was physically possible.

What Bill needed was an unequivocal display of support. It came courtesy of one of the most unsavory figures in the history of American forensic science.

As always, Albert H. Hamilton strode assertively to the stand. Although small in stature, this upstate New Yorker managed to convey an impression of immense gravitas and erudition. In a confident voice he announced that he was a criminologist of forty-seven years' standing. But that didn't begin to tell half the tale.

He had begun his career as a producer of patent medicines in Auburn, New York, late in the nineteenth century. An undeniably shrewd student of human nature, he sensed the public's growing awe of all things scientific, and realized that a title might be a useful marketing tool. Henceforth he encouraged clients to call him "Doctor," and the moniker stuck. As science moved into the courtroom, Hamilton found himself powerfully attracted to the role of "expert witness." At fifty dollars a day plus expenses, it provided a lucrative income, especially if one was prepared to exercise a certain flexibility when it came to ethical standards.

Masquerading under the title of "Micro-Chemical Investigator," he promoted himself by way of a pamphlet titled *That Man from Auburn*. This lyrical paean of self-glorification sang his praises as an unrivaled expert in chemistry, microscopy, handwriting analysis, toxicology, bloodstain identification, causes of death, fingerprinting, and anatomy. Clearly Dr. Hamilton was an expert for all forensic seasons. Still, he wasn't satisfied. After studying a few European reports on early attempts at bullet identification, he armed himself with a microscope and a camera and sallied forth as a ballistics expert for hire, convinced he could hoodwink any jury in America.

He first came unstuck in his home state of New York in 1915, when his deliberately misleading testimony led a dull-witted German immigrant, Charles Stielow, to the very brink of the electric chair. Only a last minute intervention by other, wiser ballistics experts averted a catastrophe and exposed Hamilton for the fraud that he was.*

It was a minor setback. In a huge country such as America, where newspaper trial coverage rarely crossed state borders at that time, charlatans like Hamilton had virtual *carte blanche* to practice with impunity.

In 1921 the venue was Massachusetts and the sensational Sacco-Vanzetti case. This time Hamilton testified forcibly for the defense during a motion for a retrial, until the judge caught "that man from Auburn" attempting to illegally switch gun barrels to shade the evidence in his favor.

Now Hamilton had traveled to Florida, hired to "prove" Lancaster's innocence. He began in typically flamboyant style.

To audible gasps, he produced Haden's skull in court, rolling it expertly in his hands as he spoke, pointing out to the jury that "the bullet was fired across the head and slightly backward."[16] The absence of scorching or powder burns on the surface, and the presence of subcutaneous powder fouling, led him to describe the wound as "sealed contact,"[17] by which he meant that the gun had been held so tightly against the head as to prevent any explosive gases from escaping.

Thus far Hamilton had been "on message," as far as the panel members were concerned. Soon, though, he started to waver. Squinting closely at the gun, he declared portentously that there were two human hairs still embedded in the sight (subsequent reexamination by prosecution experts revealed these to be cotton fibers). Unperturbed by this hiccup, Hamilton bulled forward, confirming that the bullet recovered from the pillow had definitely been fired from the .38 Colt.

After describing the autopsy report as the best he had ever seen, Hamilton delivered his verdict: "Absolutely suicide. There is not a scintilla of evidence to support a theory of homicide or murder."[18]

The "sealed contact," he said, was incontrovertible proof of suicide. He argued that the natural reaction of someone being touched while asleep would be to flinch away, especially from a gun barrel pressed against the head. District Attorney Vernon Hawthorne looked incredulous. "You claim

---

*See Colin Evans, *The Casebook of Forensic Detection* (New York: Wiley, 1996).

it would be impossible to hold the head of a man lying down asleep and kill him by shooting him with a pistol?"

"Impossible."[19]

Far too late in the day—and possibly to buy himself some thinking space—Hawthorne now attempted to trash Hamilton's qualifications. "How did you acquire the title of doctor?"

Hamilton just brushed him aside. "Lawyers started that. I am not a doctor."[20] And he carried this lofty disdain into his final pronouncement: "I found nothing to support anything but suicide. I say this not as an opinion, but actual knowledge."[21]

It was vintage Hamilton: preposterous, full of bluster, high-flown psuedoscientific jargon, and unverifiable "facts," yet it was oddly compelling. He wasn't in the business of ambiguities or qualified maybes; he gave the jury twenty-four-karat certainties. And judging from their approving glances as he took his leave, they plainly appreciated his forthrightness.

If Hamilton was good as a witness, then Bill was downright sensational. As Carson suspected, his client's air of open-faced honesty—whether genuine or pure artifice—created just the right impression. He was never too eager to condemn Haden; never too slow to defend Chubbie. In measured, regretful tones, he described his final conversation with Haden: the young man's suicidal depression, prompted by fears that Chubbie wouldn't marry him; insecurities over the book; and, above all, his dread that venereal disease had forever ruined his life. Bill rebuked his own thoughtlessness in leaving the revolver on the washstand, and his horror and trepidation when he found to what purpose it had been put. He took full responsibility for his "unworthy, foolish, and cowardly"[22] actions in forging the suicide notes.

Outpunched at every turn, the prosecution couldn't lay a glove on him. As for the defense, by the time Carson's consummate powers of rhetoric had run their course, Haden Clarke had taken on the mantle of Lucifer himself, castigated as a worthless scoundrel whose only decent action in life had been putting a bullet in his own brain. By contrast, Bill's utter ruthlessness—it should be remembered that, apart from his dubious conduct on the night of the killing, this was a man who had abandoned his wife and two children without a penny to pursue another woman and his own aspirations—was well and truly buried beneath a lump-in-the-throat depiction of Captain William Lancaster, honorable British officer and all-round gentleman.

On August 18, 1932, Bill was acquitted, to loud cheers in the courtroom. Chubbie, whose own courtroom performance had been understated and a bitter disappointment to the press gallery, professed herself delighted. "I knew old Bill would come through."[23]

However, she was conspicuously absent eight months later, when Lancaster set off from England, determined to break the air-speed record to Cape Town in South Africa. Loyalty was one thing, love was another. Chubbie had decided to pursue a life on her own.

Hastily contrived and poorly prepared, Lancaster's record attempt struck problems almost right away. Already behind schedule, on April 13, 1933, he and his plane disappeared somewhere over North Africa. As search parties failed to find any wreckage, rumors festered and grew. The most common had Lancaster, consumed by guilt, deliberately crashing his plane into the ocean, having the last laugh on the world by committing suicide, just like he fooled everyone into thinking Clarke had done.

The mystery of Bill Lancaster's disappearance was not solved until February 12, 1962, when a French army patrol in the Sahara found the twisted wreckage of a tiny plane and the mummified remains of its pilot. Until dying of thirst, Bill Lancaster had kept a meticulous diary of his eight-day vigil in the burning desert. In it he expressed his undying love for Chubbie. There was not a single mention of Haden. His final despairing entry—"I have no water"[24]—was dated April 20. The next day he died, one year to the day after a bullet ended the life of Haden Clarke.

So did Bill Lancaster really kill Haden Clarke? Years later, when asked that question, Chubbie would only say, "I don't know."[25] Clearly the doubts had set in. Perhaps they were there from the outset. After all, it was a very strange suicide.

In his final speech to the jury, Assistant DA Henry Jones said, "When you get out there in the jury room, lie down on the floor and see if you can shoot yourself through the head where this man did."[26]

A .38 Colt revolver is a long-barreled, heavy weapon, with a trigger pull of eight to fourteen pounds. To produce the bullet trajectory found in his own shattered skull, Clarke would have had to lie down and balance the gun against his temple; then, unless he was extraordinarily flexible, he would have to depress the trigger with his thumb. Such dexterity sounds both implausible and unlikely.

On the other hand, the bullet trajectory is exactly what one might expect had the shot been fired by someone standing between the two beds. Fired from slightly below and to the side of Haden's head as it lay on the pillow, the bullet would enter the right temple from front to back, then naturally exit above the left ear.

The defense would not be rebuffed on this point. Having successfully dragged the deceased's character through every gutter in Miami, they had the nerve to suggest that the bizarre positioning of the suicide gun had actually been Haden's last stroke of decency—ensuring that the bullet was aimed away from Bill's bed.

The argument that Clarke was displaying humanitarian concerns for his roommate scarcely stand up to scrutiny. Had he wanted to recline and shoot himself, there were two other beds in the same room, both out of the line of fire of Bill's bed. Or he could have sat on the edge of his own bed and fired at right angles to his roommate. Better still, why not, if indeed he was racked by terminal pangs of conscience, simply choose another room in the house for his act of self-destruction?

These were, of course, ridiculous suggestions, but then so much in this case bordered on the absurd. Throughout, Lancaster was treated with the softest of kid gloves. Had he been tried in a modern courtroom, he would have found himself facing a far more stringent cross-examination, with none of that hokum about British standards of "fair play" Hamilton would have been exposed for the fraud that he was, and the jury, in all likelihood, would have arrived at a different outcome.

Chapter 6

Sir Henry John Delves Broughton (1941)

Murder in High Places

At about 3:00 A.M. on January 24, 1941, two Kenyan dairy workers driving their truck on a road just outside Nairobi were suddenly dazzled by the glare of blazing headlights. As they braked and squinted through the rainy darkness, it became apparent that a car had slewed off the road ahead. Both workers ran across for a closer look. The car, a right-hand-drive Buick, had come to rest, tilting precariously, with its driver-side wheels on the edge of a ditch. At first glance the car appeared to be empty; then the two workers realized that in the footwell beneath the dashboard lay the motionless figure of a European male. Once they'd established that the man was dead, the two workers made straight for the nearby police station at Karen, a suburb of Nairobi.

After much vacillation—violent deaths of white settlers were always serious affairs in colonial Kenya—anxious local officers decided to refer the incident to a higher-up. The call went out, and at 4:50 A.M., Assistant Superintendent Anstis Bewes, a part-time policeman with little training, duly arrived at the intersection of the Ngong and Karen Roads, where the wreck had occurred. With nothing obvious to suggest otherwise, Bewes assumed he was dealing with a simple traffic accident and began making notes. As he examined the car's interior his nose wrinkled—someone had been wearing a very strong perfume. He also noticed that the ignition had been switched off, which struck him as odd, as crashed car engines usually just cut out. Moving on, he tried to make out the dead man's face, but the position of the body made this impossible.

As Bewes completed his initial report, more officers arrived, milling haphazardly around the car. Their unwitting clumsiness combined with the steady drizzle to ensure that any possible footprints or tire marks were soon obliterated. When Superintendent Arthur Poppy, the head of the Nairobi Criminal Investigation Department, was eventually phoned with details of the incident, he ordered photographs to be made of the crash, unaware of just how much degradation the scene had suffered.

Quite by chance, at 8:00 A.M. the local government pathologist, Geoffrey Timms, happened to pass by the intersection on his way to work. Timms did what he could in the cramped confines of the car. He noted that the hands and face of the dead man were cold and rigor was setting in; then he ordered that the body removed so he might examine it more closely.

Only then, laid out on a stretcher, was the dead man finally recognized. It was one of those occasions when "Good God!" seemed the only appropriate comment to make.

The origins of this tragedy could be traced to just over two months earlier and some two thousand miles to the south, when Sir Henry John Delves Broughton, a wealthy racehorse owner and big-game hunter, bagged the biggest prize of his life. After five years of intermittent, often frustrating, pursuit, he had finally succeeded in steering the delectable Diana Caldwell up the matrimonial aisle. For the fifty-seven-year-old "Jock" Broughton, who viewed most things in life through a sportsman's eye, Diana was definitely a "good catch." Not only was she thirty years his junior, but also, as one of London's top society icons, she oozed a hormonal magnetism that could turn heads and hearts with equal ease. Men found her air of icy, blond detachment sexily hypnotic; and if some of her female contemporaries thought they detected a certain pushy commonness, well, that might just have been envy.

The couple escaped wartime Britain, sailed to South Africa, and exchanged vows in Durban on November 5, 1940. It was a strange marriage from the outset. Prior to the ceremony they had signed an accord in which Broughton agreed that if Diana fell in love with a younger man, then not only would he not object to a divorce, but he would also support her for a minimum of seven years to the tune of £5,000 ($20,000) a year, a huge sum for the time. It was generosity born out of desperation. Broughton, his best years behind him and terrified by the prospect of a lonely dotage, was prepared to pay any price for Diana's company.

As for Diana, had she known just how comprehensively Broughton had squandered the family fortune—to the point of a couple of decidedly questionable insurance claims—then she might have been less enamored by the notion of marrying the eleventh baronet of Doddington.

Whatever the motives that united them, the newlyweds honeymooned in Kenya, where Broughton owned a coffee plantation. Their first port of call in Nairobi was the Muthaiga Country Club, chief watering hole for all the British ex-pats who had settled in nearby "Happy Valley."

Situated north of Nairobi in the White Highlands, Happy Valley referred more to an attitude than a geographical location. Sex, alcohol, and drugs were the lifeblood of this hedonistic paradise, where wives and husbands were swapped like baseball cards and just about every kind of carnal appetite was catered for and tolerated. In the steamy tropical atmosphere, scandal was so commonplace as to rarely rate more than a bored yawn.

Which made it the perfect lair for the predatory Josslyn Hay, twenty second Earl of Erroll, the Hereditary High Constable of Scotland, and bedroom carnivore *par excellence*. Impossibly handsome and imbued with an unassailable sense of his own superiority, the thirty-nine year-old Erroll was "King Stud" in Happy Valley. He pursued women the way other men hunted lions on safari, voraciously and always hungry for another trophy to hang on the wall.

Now that Britain was at war with Germany, Erroll had added the role of assistant military secretary for Kenya to his roster of official titles. During the 1930s, like many of his aristocratic British counterparts, he had espoused some ugly Fascist views, but the outbreak of hostilities brought an end to all that. Personal views played second fiddle to duty, and he was as patriotic as the next man.

On a daily basis the war mattered little in East Africa, certainly not enough to hinder Erroll's energetic social life. He lived on credit, drank like a demon, bedded every woman in sight, refused to drive at less than 80 mph, and had charm to burn. As the uncrowned leader of the local community, it was only a matter of time before he encountered the newly arrived Broughtons.

The fateful meeting took place in late November at a Muthaiga club ball. Diana later recalled the moment when Erroll's heavy-lidded gaze fixed itself upon her: "I had the most extraordinary feeling . . . that I was suddenly the most important thing in his life."[1]

In short order the trio became inseparable, dining regularly at Muthaiga and at the mock-Tudor house, with its fifteen servants, that Broughton had rented. At first Broughton was flattered by the attention being lav-

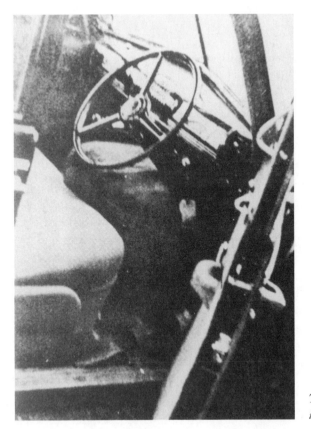

The body of Lord Erroll, stuffed into the footwell.

ished upon his attractive young wife, but by Christmas it became blindingly obvious that Erroll's interest in Diana—and hers in him—was going to provide an early test of that eccentric prenuptial pact.

Ordinarily, Erroll's trysts were discreet, hole-in-the-wall affairs, but here he was acting like some love-struck teenager. Diana responded in kind. On the dance floor they writhed in each other's arms—"Glued together as if they were making it,"[2] in the words of one local—while Diana brazenly flaunted the triple string of pearls that Erroll had given her.

The end of December saw Broughton pitchforked aside like so much chaff, reduced to brooding impotence. On the Nairobi cocktail circuit he cut a sorry sight among the gaiety and the chinking of glasses. He was never known previously as a heavy drinker, but now his alcohol consumption soared, leaving him frequently the worse for wear. Compounding his misery was a series of mocking anonymous letters addressed to him at Muthaiga, chiding him for his wife's infidelity, querying what he intended to do about the reprehensible Erroll.

Broughton churned in agony. Just about everybody in Nairobi, it seemed, knew of the torrid affair; he was being made a laughingstock. And yet, with the stereotypical British upper-class stoicism of the time, he soldiered on as if nothing had happened.

Until January 18, 1941.

That was the day when Broughton finally confronted Erroll. The latter's intransigence—he refused point-blank to stop seeing Diana—convinced Broughton that his eleven-week-old marriage was in ruins and that he should heed the advice of his friend Jack Soames and "cut his losses"[3] like a good betting man. In characteristic fashion he affirmed his intention of honoring the pact and agreed to settle £5,000 on Diana.

Broughton recalled the conversation as quiet and dignified. "He [Erroll] did say he felt miserably unhappy about the whole thing, as we had been such great friends. Both of us were as dispassionate as possible."[4]

Erroll, too, commented on this meeting, telling a friend: "Jock could not have been nicer,"[5] adding thoughtfully, "As a matter of fact, he has been so nice it smells bad."[6]

This brings us to the night of January 23, and yet another of those interminable dinners at Muthaiga.

In public, at least, Broughton displayed little evidence of any resentment he may have harbored toward Diana and Erroll, and on this particular night he was magnanimity personified. With the champagne corks popping like firecrackers, Broughton generously raised his glass to toast the glittering couple: "I wish them every happiness, and may their union be blessed with an heir."[7] Nor did he make any objection when Erroll suggested to Diana that they go dancing at the nearby Claremont Club, though he specifically requested his wife to return home by 3:00 A.M.

After the couple departed, it was a very different story. As Broughton continued to gulp down cocktails, his mood blackening with every swallow, slowly the veneer of *mari complaisant* began to peel away. One guest heard him growl, "I am not going to give her £5,000 and the Karen house. She can bloody well go and live with Joss [Erroll]. We have only been married a few months, and look how it is for me."[8]

At 1:30 A.M., stumbling drunk, Broughton was bundled into a car and driven home. With him was June Carberry, a close friend of Diana's. She and the housekeeper, Dorothy Wilks, helped Broughton upstairs to his bedroom; then June retired to a guest room. As she was readying herself for bed, she was startled to see Broughton, clad in a dressing gown, standing at her door. He merely asked if she was feeling better. June, who had complained earlier of a touch of malaria, said yes, and Broughton left. The time, so June estimated, was 2:10 A.M.

Fifteen minutes or so later, a commotion downstairs—the slamming of car doors and loud voices in the hallway—heralded the return of Diana and Erroll. Another ten minutes elapsed before June heard Erroll's car speed off into the night, whereupon Diana came upstairs to June's bedroom and the two women chatted excitedly until 3:00 A.M., when Diana retired.

June Carberry's bedroom seems to have been something of a transit camp that night, as half an hour later Broughton reappeared. For someone ostensibly teetering on the brink of an alcohol-induced coma, he was proving remarkably wakeful. Again he inquired after June's welfare, again he was assured that everything was fine, and again he disappeared. The only other sound to disturb that troubled night was Diana's dachshund. June heard it barking at some point, though she was uncertain whether it was before or after Broughton's second visit.

Finally, at approximately 3:30 A.M., silence settled over the house.

Body Discovered

The call came at nine o'clock the next morning. June Carberry reached groggily for the receiver. It was a family friend with unbelievable news— Lord Erroll had broken his neck in an auto accident!

In a state of shock, June shouted for Broughton. He came running. When told what had happened, he gasped, "Good God!"[9] and slumped onto the bed.

Within minutes a police car pulled into the driveway. In such a closed community, local officers knew who to contact in an emergency. Even before they were out of the car, Broughton came hurrying toward them, crying frantically, "Is he all right? Is he all right?"[10]

The officers confirmed that Erroll was, indeed, dead, the victim of a tragic accident. Inside the house it was bedlam. Diana was hysterical and close to collapse, unable to speak; Broughton was scarcely any better. Unable to get much sense out of anyone present, the officers left. No sooner had they gone, than June sped off to Erroll's house, intent on recovering a bundle of Diana's love letters before they fell into police hands.

Happy Valley was taking care of its own.

Meanwhile, at the mortuary in Nairobi, all sorts of investigative goodies were being uncovered. Originally, a glob of heavily congealed blood on the side of Erroll's head near the left ear was thought to have been caused by the metal spike of the headlight switch, which was without its knob. However, when swabbed clean, the injury told a different tale. A ragged fringe of black powder burns around the wound could have no other interpretation—Lord Erroll had been shot at close range.

Broughton's actions that first morning were mystifying. He burst into Karen police station, clutching a handkerchief in his tremulous hands, begging to be allowed to place it on the body. He cried, "My wife was very much in love with Lord Erroll."[11] He was directed to the mortuary, and although he was not allowed to see the body, the handkerchief was taken and placed, as requested, on the dead man's breast.

From there, Broughton drove into Nairobi and to the offices of the Union Castle shipping line. Some days earlier he had canceled an intended passage to Ceylon with Diana. Now he rebooked his ticket. Then he returned to his house, ordered his head servant, Abdullah bin Ahmed, to get a can of gasoline he had stashed a few days earlier, and meet him at the garbage pit.

Sir Jock Broughton was about to have a bonfire.

Superintendent Poppy was in a grim mood. The absence of any gun in the car had vaulted Erroll's "accident" into the category of murder, and already it was apparent that the crime scene had been hopelessly compromised. In a sour mood, he oversaw a belated examination of the Buick.

They found a cigarette and a hairpin, both stained with blood. There also were splashes of blood on the windshield and on the front passenger seat, where Erroll's head must have fallen. Judging from the presence of blood on his trouser leg, he had bled in an upright position at least long enough for the blood to drip down. This tallied with later notions that the killer must have shoved Erroll under the dashboard to steer the car into the ditch. In the back of the car the carpet was scrunched up and marked with what looked like shoe whitener, such as might be used on sneakers. Similar marks were found on the backseat.

When government scientist Dr. Francis Vint attempted to reconstruct the shooting, he started by examining the wound to Erroll's head. From this he concluded that the gun had been fired from no more than eighteen inches away, a generous estimate given the intense powder burns evident in photographs presented at the trial. Since Erroll was shot from the left, this placed the killer either in the passenger seat, standing on the running board, or leaning in through the open doorway if the car was stationary. Two factors suggested that a first shot had missed completely. There was a bruise on Erroll's head, consistent with him having struck himself on the steering wheel as he ducked to avoid the .32-caliber slug that had ricocheted off the central door pillar at about head height on the driver's side and landed by the gas pedal. He wasn't as lucky next time. A second .32 slug penetrated his skull in a direct line from ear to ear, lodged in his medulla, and killed him instantly.

The experts had told Poppy how the crime was probably committed. Now he needed to put a face behind the gun.

As he soon discovered, there was an embarrassment of suspects. The list of cuckolded husbands and jealous ex-lovers who might have wanted Erroll dead stretched from one end of Happy Valley to the other. But getting them to talk was a hellish task. Closing ranks was a way of life in this tightly knit, incestuous community, and they didn't take kindly to outsiders, particularly nosy police officers, digging into any dark corners. Insulated by wealth and privilege from the strictures of conventional society, Happy Valley adopted an air of knowing condescension toward the ex-London copper, a sense that he was a tiresome little chap who really ought to leave them alone. Besides, deep down, many felt that the libidinous earl had finally gotten his just desserts, and they were disinclined to aid any murder investigation. Who knew what other dirty little secrets it might uncover?

Despite this caginess, Broughton soon emerged as the prime suspect. He certainly had motive. Being the registered owner of numerous weapons gave him the means. But did he have the opportunity?

It all hinged on time. According to June Carberry, she had seen Broughton twice in the early hours, at 2:10 A.M. and again eighty minutes later. This meant that at the moment when Erroll was murdered—between 2:35 A.M. and 3:00 A.M.—Broughton's whereabouts were unknown. Was it possible, Poppy mused, for Broughton to have made his way through the night to the intersection where Erroll was shot, a distance of 2.4 miles away, and back again to the house within the allotted time? And were those strange visits to June Carberry's room really an attempt to establish an alibi? They were things to ponder.

Because of the heat, funerals tend to be held quickly on the equator, and on January 25 Erroll was laid to rest. That afternoon Poppy went to Broughton's house, and from servants learned of the master's hasty bonfire on the day of the murder. It was the first time Broughton had been known to do such a thing. After sprinkling gasoline liberally onto a mound of garbage, he had tossed in a match, then watched the blaze for a while. When the flames threatened to spread out of the pit, he had ordered Abdullah to douse the fire. Then Broughton retired for lunch.

With the tropical night closing in fast, Poppy delayed a search of the pit until the following day. The next morning, detectives sifting through the charred remains recovered a partially burned golf sock.* On it was what appeared to be a bloodstain.

By now Poppy was convinced that Broughton was intimately involved in the death of Erroll. Broughton's erratic behavior, both on the night of the murder and in its aftermath, and the suspicious bonfire, added up to a picture of a man with something to hide. The screws tightened still further when a check of firearms certificates revealed Broughton as the recorded owner of at least two revolvers, both Colts, one .45-caliber and the other .32. It was the latter gun—the same caliber as the murder weapon—that intrigued Poppy. But when he asked Broughton to produce the gun for testing, he was told that it had been stolen just three days before the murder.

Sure enough, on January 21, Broughton had reported a burglary at his home. The thieves, he said, had taken a small sum of money, a silver cigarette case, and two revolvers, including the .32 Colt, from the study. At the time Broughton had explained to puzzled officers, who were unable to find any signs of a break-in, that he must have left the veranda door unlocked, though he was less forthcoming about how the thieves had managed to avoid the security alarm bell hidden under the carpet and that would have rung had any unauthorized person entered the living room. After routinely logging the theft, the officers left.

Convinced that the burglary had been faked, Poppy widened the investigation. In an interview, Diana let slip that on a recent visit to Jack Soames's farm, Broughton had spent hours practicing his revolver shooting. When two detectives visited the farm at Nanyuki and raked over the ground where the targets had been erected, they soon found several spent shell casings and some live rounds from a .32 revolver. These were sent to government scientists for comparison with the two bullets from the murder scene.

Ever since the sixteenth century, armsmakers have known that by etching a spiral groove into the gun barrel, spin is imparted to the projectile, giving it much greater accuracy. It is this groove—or "rifling"—that leaves the distinctive marks known as striations on the bullet itself, and that forms the bedrock of modern ballistics study. The raised parts between these grooves are known as "lands."

---

*Although Broughton would later deny ever owning such socks, it emerged that Paula Long, a famed local beauty, remembered him wearing them all the time, and had a photograph to prove it. Significantly, she never showed this to the police—yet another example of the Happy Valley obstructionism that plagued this case.

The identification of a bullet as having been fired from a particular weapon is possible primarily because of what occurs during manufacture of the barrel. First, it is smooth-bored, then reamed to a specified diameter, before being rifled. Because the tools used to make a barrel wear down minutely with each succeeding gun, it is impossible to produce two identical rifled barrels. Bullets fired from different guns will always have different striations. By the same token, each barrel retains unique characteristics, which it will impart to every bullet it fires.

After six weeks of painstaking analysis, the report came back—the bullet that killed Erroll and the bullets recovered from Nanyuki had undoubtedly been fired from the same gun. This was the clincher for Poppy: Broughton had obviously faked the robbery of his two Colt revolvers, while laying the groundwork for a premeditated murder.

On the strength of this ballistics evidence—and, it must be said, very little else—Broughton was arrested on March 10 and charged with the murder of Lord Erroll. "You've made a big mistake,"[12] he told Poppy, then added, "Do you mind if I have a whisky?"[13] Police methods in colonial Africa being somewhat more relaxed than those elsewhere, Poppy immediately gave the accused man a reassuring shot from his own hip flask.

While Broughton languished in jail awaiting trial, Diana had been busy. Although her affair with Erroll had been scandalously public, the rather bizarre etiquette of Happy Valley now demanded a show of solidarity with her beleaguered husband, and this she provided in spades. Her greatest service was to hire the celebrated South African lawyer Harry Morris, a bruising cross-examiner and masterly tactician, to defend him. Morris didn't come cheap—his fee was £5,000—but he was first-rate and, vitally, he had an encyclopedic knowledge of firearms, enough for him to confidently predict that he could defeat the prosecution's case on one simple point of ballistics alone, though he adamantly refused to tell anyone, including his client, what this case-breaker would be.

Jealous Defendant

The trial began on May 26, and on each day of the five weeks it lasted, the court heaved with spectators panting for every salacious detail of life, love, and murder in Happy Valley. As expected, the Crown relied heavily on circumstantial evidence. Prosecutor Walter Harragin depicted the defendant as a man torn by jealousy, driven to murder by humiliation, someone who, after saying good night to June Carberry at 2:10 A.M., had slipped out of the house, committed murder, and then slipped back again, ensuring that June again saw him and would provide his alibi.

Anticipating defense skepticism about the likelihood of a fifty-seven-year-old man jogging five miles round trip through muddy fields in the middle of the night to commit murder, Harragin fine-tuned this theory, suggesting that Broughton may have either hidden in the backseat of Erroll's car or else hitched a ride from his intended victim before shooting him at the crossroads, then running back across the fields to the house.

Fine as far as it went, except that Harragin was never able to explain satisfactorily how Broughton managed to slip in and out of the house undetected. The stairs were notoriously creaky, and it was hard to picture a man, far from the first blush of youth, shinning up and down a drainpipe to his second-floor bedroom.

Nor was Harragin able to overcome the awkward fact that for Broughton's alibi to work, the body had to be discovered early. Had it lain undetected for another couple of hours, then no amount of nocturnal assignations along the landing would have assisted Broughton.*

Wisely, Harragin moved on, highlighting Broughton's previously unsuspected interest in yard fires, his sudden rebooking of the passage to Ceylon, and a string of very suspicious questions and remarks made during police interviews. For instance, Broughton had asked Poppy if a man could be hanged for shooting his wife's lover after catching them in flagrante delicto, and on another occasion he had asked what the chances were of finding a gun "buried somewhere in Africa."[14]

Ultimately, the only direct evidence linking Broughton to the crime scene—the only evidence that could hang him—was the .32 slug that had been removed from Erroll's brain. Linking this bullet to Broughton's missing gun was central to the prosecution's case. They needed to prove that the bullets found at Nanyuki matched the murder bullets, with the overwhelming probability that all had been fired by the Colt .32, registered in Broughton's name and allegedly stolen three days before the murder.

To do that the prosecution needed their expert witnesses to come up to proof. The chief government chemist, Maurice Fox, suffered a shaky start with his admission that he had been unable to positively identify the stain on the charred woolen fragment found in the pit as blood, but when it came to the bullets, he caught his stride. All of them, he said, had been fired from one weapon, stating, "This a conclusion, not an opinion."[15] He also

---

*Significantly, Broughton claimed to have no recollection of either visit to June's bedroom, bolstering the belief that he had drunk to the point of insensibility and was therefore incapable of such murderous athleticism.

confirmed that live rounds found at Nanyuki contained the same pre–World War I black powder that had left the powder burns on Erroll's head.

For almost eight hours he bombarded the jury with a perplexing array of comparison photographs taken of the two crime scene bullets and the four slugs recovered from the impromptu shooting range at Nanyuki. All, he said, had been fired from a weapon that had five right-hand grooves. With the aid of photographs, he pointed out a sequence of some twenty-six points of similarity in all six bullets, which, he said, put the matter beyond doubt.

"Is there any possibility of a mistake in your conclusion?" Lord Chief Justice, Sir Joseph Sheridan asked.

"No,"[16] Fox replied categorically.

Morris always believed that expert witnesses—with their polished certainty and air of invincible self-esteem—were often the easiest prey for a skilled cross-examiner, and the man from Johannesburg was matchless in the fine art of needling witnesses into evidentiary indiscretions. Fox, truculent and didactic, took up the challenge. After conceding that he was not a ballistics expert *per se*, he airily added, "I look upon an expert as a person who knows everything about nothing."[17] Witty paradoxes of the kind employed by Oscar Wilde might set theater audiences rocking with laughter, but they rarely succeed in the courtroom, as Wilde found to his own cost. Now it would be Fox's turn to discover how easy it is for witness stand cleverness to generate jury box antagonism.

Slowly and skillfully, Morris baited the trap. He professed himself unable to see the photographic similarities between the bullets that Fox found so obvious and requested enlightenment from the witness. As Fox droned on, a torpor settled over the court and especially the jury, until abruptly Morris went on the attack. Thus far, he said, the witness had mentioned only the similarities between the bullets. What about the *differences?* As Morris well knew, even the same gun produces minute differences between two bullets fired from its barrel.

Fox spluttered and blustered under Morris's relentless insistence that without the firearm itself—the .32 Colt that the Crown could not produce—a definitive identification was next to impossible. To support his argument, Morris quoted from a textbook written by Sir Gerald Burrard,* one of the great ballistics pioneers.

Again hubris got the better of Fox. "I am quite ready to agree that Mr. Burrard knows a considerable amount," he said. "I think I know as much about the characteristics reproduced on bullets from firearms."

The Identification of Firearms and Forensic Ballistics (New York: A. S. Barnes, 1962).

"Perhaps more?" goaded Morris.

"Who knows?"

After more sparring, Morris brought matters to a neat close. "So long as you contradict Burrard, as read here, I am satisfied."

"So am I,"[18] snapped Fox, peevish to the last.

Such absolute certainty as that displayed by Fox often has a negative effect on a jury. To say "I am right and all those who differ from me are wrong" is apt to generate feelings of hostility in the minds of those only too conscious of their own fallibility.

Morris moved in for the kill. Throughout the trial the prosecution had placed great weight on the fact that all six bullets had been fired from a revolver whose barrel had five grooves in the rifling, which twisted along the barrel in a right-handed direction. Morris took up this theme with Ernest Harwich, the other government ballistics expert. "In all these bullets, was the direction uniform?" he asked.

"Yes," said Harwich. "It is right hand in all the bullets."

"Is the direction in a Colt revolver right or left?"

Harwich cowered like a rabbit caught in a car's headlights. He knew what was coming. Clearing his suddenly parched throat, he croaked, "Left in the barrel." Morris pressed home his advantage. "Can you say what kind of a gun the bullets came from?"

"I can say they came from a revolver."

At this point the judge interjected: "But not a Colt?"

Poor Harwich could only reply miserably, "As far as my experience goes, all Colt revolvers have *six* grooves and a *left*-hand twist [emphasis added]."[19]

At long last Morris had fired his brief-busting howitzer, the single issue that, he believed, would blow the prosecution's case out of the water—on the Crown witness's own admission, Broughton's missing .32 Colt could not possibly have been the murder weapon!

This was devastating. How the Crown could have erred quite so spectacularly beggars belief. Prosecution blunders are gold dust to any good defense counsel, and in Morris's hands, this gaffe loomed larger than Mount Kilamanjaro. Why, he thundered to the jury, if the "experts" made such a fundamental flaw in this aspect of their evidence, should you now believe a single word of their testimony?

Such was the scale of this blunder that it entirely overshadowed one important fact: It did nothing to eliminate Broughton as the killer. Who was to say he had not shot Erroll with another .32 revolver? But as Morris was quick to point out, no such weapon had been produced, and no evidence of its existence had been found.

Eager to paper over any cracks before they might appear, the defense called its own firearms expert. Captain Thomas Overton, Assistant Inspector of Armorers to the East African Forces, had spent most of his twenty-nine years in the army in the identification of bullets, and had testified in numerous cases. In his opinion the comparison photographs were useless, and he'd found enough discrepancies to suggest that the two crime scene bullets were fired from one weapon, and the four bullets found at the shooting range from another.

Piece by excruciating piece, Morris dismantled the evidence of Fox and Harwich until all that was left was a pile of rubble. He lambasted their wasted months in the laboratory, their detailed comparative photographs, which he dismissed as too confusing to be comprehensible to the jury, and as for their conclusions about the weapon that had fired the bullets. . . .

By now the prosecution was in total disarray. All their hopes had been pinned on Broughton's missing .32 Colt, and it had blown up in their faces. Some indication of their desperation can be gauged from the fact that when Fox was recalled, he now claimed to have found *thirty* points of similarity between the bullets. Like many an expert witness before and since, Fox seemed congenitally incapable of admitting an error, and it showed in his final dogmatic rebuttal.

"Having heard various criticisms of the photographs, have you any doubts that the bullets at Nanyuki and the crime bullets were fired from the same gun?" asked Harragin.

"No, none."[20]

But it was too late. Mistakes linger long in the mind of a jury, and there was nothing in Broughton's evidence to upset their calculations. During five arduous days on the stand, he repelled with majestic ease everything the prosecution threw at him.

It was said at the time that no jury of white men (as they then were in Kenya) would ever have convicted Broughton or anyone else of murdering Erroll, as the latter was said to have slept with most of their wives. And so it proved. On the morning of July 1, as the jury returned with their verdict, the foreman winked at Broughton and gave him the thumbs-up.

Not guilty.

Broughton laughed his way out of the courtroom. He had escaped the gallows by a hairbreadth, but he could do nothing to wipe off the mud that stuck to his name, as he found himself ostracized by former friends, and shunned by Diana. She had performed her duty; now it was over.

In November 1942 Broughton returned alone to the land of his birth. Even in wartime and across thousands of miles, his notoriety had preceded him, and there were some awkward questions from both family and police about trust fund money that had gone missing a few years earlier.

It was all too much. A few weeks later, on December 2, doomed to social banishment and possibly a prison term as well, Broughton booked into the Adelphi Hotel in Liverpool, with orders not to be disturbed. Two days later he was found comatose in his room. He had injected himself fourteen times with barbiturates. He died the next day.

Other Suspects

So who did kill Lord Erroll? Dozens have claimed to know the "real story," but versions vary according to the source. According to Juanita Carberry, fifteen-year-old stepdaughter of June, Broughton confessed to her just one day after the killing. And she had another story to tell. She had seen the infamous bonfire and distinctly remembered seeing a pair of sneakers among the smoldering rubbish, which was highly unusual, because in Kenya old shoes were invariably passed on to servants. (Again this information was kept from the police.) If true, it does raise the tantalizing possibility that these sneakers might have been the source of the obscure shoe-whitener marks in the Buick.

Another rumor claims that one of Erroll's former lovers, a tempestuous American heiress named Alice de Trafford, also admitted to the killing. But perhaps the most intriguing "solution" came courtesy of someone much closer to the action than most: Dorothy Wilks, the housekeeper at Karen. Off the record, she confided her belief that Diana was the guilty party. In Wilks's version, Broughton, far from agreeing to honor his prenuptial obligations, had actually welshed on the agreement, and Erroll, once he realized that Diana would not come bearing a £5,000 per year dowry, rapidly lost interest.

Curiously, this view was supported by Isie Maisels, a local lawyer who was originally offered the Broughton brief, which he had to decline because of military commitments. Speaking years later, Maisels recalled attending a luncheon with the Carberrys. At the end of the meal, Lord Carberry said: "Let's get down to business. . . . Well as you know, girls will be girls. . . . Look, I want you to tell me what Junie [his wife] should say in evidence."[21]

Maisels hurriedly explained that it was not part of counsel's work and duty to tell a witness what to say; indeed, such conduct was highly improper. But the comment "girls will be girls" stayed with him, and strengthened his suspicion that Erroll had been murdered by Diana, possibly aided by June Carberry.

One fantastic story, recently published,* argues that the real killer was a British secret agent, under orders to assassinate Erroll because of his known Fascist sympathies. Without a scrap of corroborating proof, beyond the flimsy memoirs of a now dead ex-Kenya colonial, the author would have us believe that the executioner was a woman whom Erroll knew intimately. With her true identity disguised by a makeup artist from the Nairobi amateur dramatic club, the woman apparently flagged Erroll down on the lonely road and asked him for a ride into Nairobi. Erroll, totally bamboozled by the disguise, had invited the woman into his car and was duly assassinated. Such bizarre flights of imagination are scarcely worthy of comment.

In the end it all boils down to credibility and sources. According to his biographer,† Harry Morris was convinced of his client's innocence, whereas Diana always maintained that Morris believed Broughton to be guilty. Interestingly, whenever Morris was asked by friends whether Broughton, notwithstanding his acquittal, had actually done the deed, the canny old lawyer would chuckle, "You know, dammit, I forgot to ask him!"[22]

---

*Errol Tyzebinski, *Dead Reckoning* (London: Fourth Estate, 2000).
†Benjamin Bennett, Harold Timmins, *Genius for the Defense* (Cape Town: Harold Timmins, 1959).

Chapter 7
Alfred De Marigny (1943)
The Bogus Fingerprint

*F*rame-up! Probably the ugliest term in criminology. Even hardened lawbreakers balk at committing a crime and deliberately piling the blame onto a wholly innocent third party. But when those doing the framing are those charged with investigating the crime, then we really are wallowing in murky waters. Fabricating evidence to achieve a conviction is bad enough. Stir in a potential death row outcome and it gets even worse. For most right-minded people the notion that someone would be prepared to "solve" one murder by organizing another is repugnant beyond belief. Yet it happens. Fortunately, the all-hands-to-the-tiller frame-up is a scarce beast, and probably will get scarcer still as media scrutiny tightens up and science and technology continue their relentless march. But that won't stop people from trying.

It was a wild night in Nassau. The storm swept in, bending the palm trees like rubber and turning the sky into a crazy paving of purple lightning; but as dawn rose on Thursday, July 8, 1943, the worst of the weather had passed, and there was the promise of yet another glorious tropical day in the Bahamas. At 7:00 A.M. Harold Christie, a forty-seven-year-old British-born property developer, emerged from his guest bedroom and onto the balcony at Westbourne, the sprawling mansion near Cable Beach that was home to the island's wealthiest resident, Sir Harry Oakes.

Oakes had hosted a small dinner party the previous evening at Westbourne at which the conversation was business, business, business. Reputed to be one of the richest men alive—he had a tax-free income of $40,000 *a day*—Oakes spoke of little else, and Christie, as ever, had been trying to interest the tycoon in his latest plans. Both men were leading lights in the so-called Bay Street Boys, a shadowy cabal of white business-men who ran local commerce and politics with an iron fist. At 11:00 P.M., when the other guests left, Christie had decided to stop over for the night and stayed in a guest room. During the night he had been troubled by mosquitoes and an occasional clap of thunder, but otherwise he had slept undisturbed.

Yawning and stretching, Christie strolled along the balcony until he reached the master bedroom, which was next to his own. It was custom-ary whenever he stayed at Westbourne for Oakes and him to breakfast to-gether al fresco. He called out, "Hi, Harry."[1] There was no reply. After a few seconds he stepped inside.

Christie froze at the sight that greeted him. Something or someone had turned Harry Oakes's bed into a grotesque funeral pyre. He lay on his back, his face blistered and blackened by soot from the smoke that rose from the still-smoldering mattress. His pajamas had been burned off, and feathers from a torn pillow stuck to his charred skin, fluttering like moths in the breeze generated by a bedside fan. A trail of scorch marks ran across the carpet and up the bedroom door. To the left of the bed stood a tall decora-tive Chinese screen, burned and spattered with blood.

Christie rushed forward. He grabbed a flask from the bedside table and attempted to revive his friend with a glass of water, but the ugly motif of wounds behind the left ear made it clear that Oakes was beyond human assistance. Christie abandoned his attempts at resuscitation and phoned his brother, telling him to get a doctor. Then he made a string of other calls to island notables, informing them of the tragedy. His final call was to the Governor General, the Duke of Windsor.

Three years had elapsed since the duke arrived in Nassau to take up his official post. Since renouncing the crown in 1936, to be with "the woman I love,"[2] the former King Edward VIII had been a huge embarrass-ment to the British government, publicly fawning over Hitler, embracing Nazism, and generally making a nuisance of himself. So it was with some relief that, when war broke out, Prime Minister Winston Churchill was able to shunt the troublesome duke off to this remote outpost of the em-pire. Nothing significant ever happened in Nassau. Until this morning.

The duke acted quickly. His first action was to embargo all news of the tragedy, under his wartime censorship powers. Then he did something strange. Rather than allow the local police force to investigate the case, he reached for the phone and called not the FBI, or as one might have expected, Scotland Yard, but the Miami Police Department. He asked to speak with Captain Edward Melchen, a homicide detective who had doubled as a bodyguard for the duke on his frequent visits to Miami. In guarded fashion, the duke requested Melchen to come to Nassau to investigate a prominent citizen, who had died "under extraordinary circumstances."[3]

Unbeknownst to the duke, his news embargo had already been breached. Earlier that morning the phone had rung at Westbourne. It was a local journalist, Etienne Dupuch, calling to confirm an interview he had arranged with Oakes. Hysterically, Christie shouted down the line, "He's dead! He's dead! . . . He has been shot."

"Shot? Are you serious?"

"Of course I'm serious. I've just discovered him. He's dead."

Dupuch gulped hard. "This is a very big news story, Mr. Christie, and I'm a journalist. I propose to cable it around the world. You *are* certain?"

"I'm positive," Christie said wearily.[4]

Despite the Duke of Windsor's best efforts, in a matter of hours the whole world knew that Sir Harry Oakes was dead.

Sixty-eight years old at the time of his death, built like a bull, and just

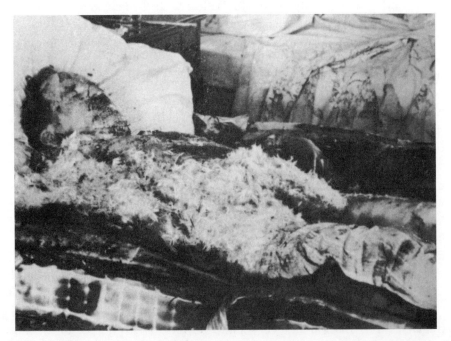

The charred remains of Sir Harry Oakes.

as ferocious, Oakes was not a likable man. In many respects he belonged to an earlier, freebooting age, when manners and morals mattered less than ambition. Born the son of a schoolteacher in Sangerville, Maine, he was bitten by the gold bug at an early age, and in 1896 he abandoned his medical studies and headed for the prospecting camps of the Canadian North, intent on carving out a fortune. To make ends meet he scrabbled for work as an orderly, treating frostbite and gangrene and malnutrition, all the while gleaning every scrap of information he could from the grizzled panhandlers, dreaming of the big strike.

For fourteen years he followed the restless waves of prospectors, chasing every elusive strike from California to the Yukon, to Australia and Africa. There were times when he almost froze to death, others when he had to dodge rattlers in order to stay alive. But he never quit. His dogged persistence paid off in 1910 when, with a partner, he finally struck gold at Kirkland Lake in eastern Ontario. It became the second largest gold find in North America and over the next dozen years generated fabulous wealth, enough to turn the hard-bitten ex-miner into one of the richest men alive.

To strengthen his links with the country that had provided his massive prosperity, Oakes renounced his American citizenship and became a naturalized Canadian. But he soon became disenchanted with his adopted country. Oakes hated taxes worse than poison, and when he found himself shelling out $3 million annually to the Canadian revenue he began searching for somewhere else to call home.

It was while wintering with his wife and five children at his Palm Beach home that he met Harold Christie. Christie, a world-class salesman, was on the lookout for rich people to buy parcels of land in the British-run Bahamas, and they didn't come much richer than Harry Oakes. Estimates of his wealth varied, but most put the figure at about $300 million. Oakes's eyes glinted as Christie boasted of life in the Bahamas. With no income tax or death duties, it sounded like paradise on earth to the acquisitive multimillionaire. In 1934 he upped roots and moved everything, especially his money, to Nassau on New Providence, largest of the seven hundred islands that make up the Bahamas. He became a generous benefactor, pouring millions into the islands, buying hotels and landscaping golf courses, funding charities to provide milk for children and hospitals for the poor.

His munificence soon spilled over into Britain itself, with one London hospital receiving a bequest of £50,000 ($200,000). Charity on this scale bought its reward. In 1939 a grateful King George VI, the Duke of Windsor's younger brother who had succeeded to the throne after the abdication, conferred a baronetcy on the chunky ex-prospector.

Sir Harry Oakes not only loved money and titles, he also craved power.

And when the Windsors arrived as the king's representatives in 1940, he wasted no time ingratiating himself. The duke, always a toady in the presence of megawealth, conferred closely with Oakes on every decision affecting the island. Christie, too, was part of this tightly knit enclave. Among them, these three men ran Nassau like a fiefdom. Oakes provided the financial muscle; Christie had the presentational skills and a bottomless well of contacts; while the Duke of Windsor was the rubber stamp that made everything possible.

Until the morning of July 8, 1943.

Imported Sleuths Arrive

Just after lunchtime that same day the Pan American Airways noon flight from Miami touched down in Nassau. On board was Captain Melchen, accompanied by another detective, Captain James Barker, an officer with a checkered career but who was now styling himself a fingerprint expert.

The two detectives were whisked off to Westbourne. Only then did they learn that the subject of their inquiries was the internationally renowned Sir Harry Oakes. Melchen was flummoxed. The duke's phone call had led him to believe that he would be investigating a suicide, but one glance at the blistered body with its four head wounds left no doubt that this was the most savage of murders.

Judging from a host of muddy footprints on the stairs, someone had broken into the house, killed Oakes, then attempted to destroy all traces of the crime by dousing the body with gasoline and setting it on fire. Except that the body hadn't burned; it had just smoldered.

The first doctor on the scene, Hugh Quackenbush, put the time of death at two and a half to five hours earlier. He couldn't hazard a closer guess because both body and bed had been burned, thereby affecting what is usually the most reliable guide, body temperature. He did spot one curious feature. A line of dried blood ran from the wound toward the top of the head. "This would indicate that the body was face down," he said later, "or it had been moved, because blood would not naturally flow uphill."[5]

The autopsy was conducted by Dr. Laurence Whilley Fitzmaurice, Acting Chief Medical Officer of the Bahamas. He found four shallow wounds, none deeper than an inch, behind the left ear, all "somewhat triangular in shape,"[6] with the apex in each pointing toward the front of Sir Harry's head. In his opinion, death was caused by shock, by hemorrhaging around the brain, and fracturing of the skull. The wounds had been inflicted "by a heavy blunt instrument with a well-defined edge,"[7] though he had no idea what the weapon could be.

Immediately after the autopsy, Oakes's body was put on a plane bound for the United States, only for the plane to be summoned back to Nassau just after takeoff. It had been learned that photographs of Oakes's fingerprints had smudged, and it would be necessary to reshoot the photos for comparison purposes. Such sloppiness typified the entire investigation.

Already the crime scene was a shambles. No area had been sealed off. Ghoulish visitors and wide-eyed reporters had been allowed to wander through unsupervised, leaving their fingerprints on chairs, lamps, windows, and doors. As a result, harvesting of trace evidence at the crime scene was severely compromised.

Not that it seemed to matter. For even before the body of Sir Harry Oakes was cold, everyone connected with the case, and most of Nassau's ruling elite, were convinced beyond any shadow of a doubt that they knew the identity of the killer.

Alfred de Marigny appears to have been one of those louche types whom others instinctively distrust. He was born on the Indian Ocean island of Mauritius, of French descent, and had inherited a languid charm that could grate or placate, depending on the observer. A champion yachtsman, he was twice married and twice divorced, and had breezed into Nassau on the back of a whopping settlement from his second wife. With his flashy good looks, the lean and lanky sailor cut quite a dash on the cocktail circuit, especially with the ladies. Vague links to French nobility added to his allure, but in this incestuous outpost of the British Empire, where one's pedigree was scrutinized with the intensity of a Kentucky bloodstock agent reviewing foals at the Keeneland sales, de Marigny was marked down as "unsound." He later attributed this animosity to his valiant efforts to improve the lot of local blacks, as they struggled to slough off three hundred years of oppressive British colonialism. Indeed, if we are to believe his remorselessly self-serving autobiography,* de Marigny was a saintly soul, programmed by nature to align himself with just about every underdog in the Western world. Evidently such altruism came at a high personal cost, to judge from the way in which each business venture he touched in Nassau careered from disaster to calamity.

De Marigny was unfazed. Even as his funds dwindled, the thirty-two-year-old playboy had his eye on the greatest prize of all—Oakes's gorgeous daughter, Nancy. There was only one stumbling block—at just seventeen years of age, Nancy was legally still a minor, bound by law to her father's wishes, and the old man despised de Marigny.

Conspiracy of Crowns (with Mickey Herskovitz) (New York: Bantam, 1988).

Oakes, the multimillionaire gold miner, could spot a twenty-four-karat gold digger a mile off—or so he thought—and made his disapproval crudely obvious. De Marigny bided his time. Just two days after Nancy's eighteenth birthday in May 1942, when she no longer needed her father's consent, the couple were married in New York.

No one could accuse Oakes of being a bad loser. Elbowing personal prejudice aside, he reluctantly accepted de Marigny into the family, even offered him a job—declined—and gave the couple a check for $5,000 as a Christmas present. But nothing could stifle his abiding distrust of de Marigny, and before long the two were at each other's throats like pitbulls. Soon Bay Street was humming with rumors that Oakes was even prepared to disinherit his daughter if she didn't ditch this mountebank. And then Sir Harry Oakes was found dead.

The absence of any signs of robbery convinced investigators that this crime was a "domestic." Someone very close to Oakes had wanted him dead. And with Oakes's wife and children—including Nancy—all either in the United States or Canada, that only left de Marigny.

On their first afternoon in Nassau, Barker and Melchen summoned the son-in-law to Westbourne. He was truthful about the antagonism between himself and Oakes, but said they had not spoken since March 30. When Melchen asked him to account for his whereabouts the previous night, de Marigny explained that, with his wife away in America receiving medical treatment, he had thrown a dinner party for some friends. At 1:00 A.M., when the party broke up, he had gallantly driven two female guests through the storm to their Cable Beach cottage, a journey that took him past Westbourne. Glancing over at the house, he noticed that the upstairs lights were still on. Then he had driven on.

Melchen and Barker weren't convinced by de Marigny's glib plausibility. Working on the assumption that whoever carried out this crime would have suffered some degree of collateral burning, they took a microscope and subjected de Marigny to an intense personal examination.

"The hair is singed on your hands and beard," said Barker. "Do you have an explanation for that?"[8]

De Marigny offered a multitude of reasons: He was a heavy cigar smoker; he regularly stoked fires at the chicken farm that he ran; but much the likeliest cause was that during the previous night's storm, power to his house had been knocked out and he'd lit several hurricane lamps to provide illumination for his dinner guests. While doing so he had burned his hand.

Still unimpressed, the detectives accompanied de Marigny to his home, where they demanded to see the clothes he had been wearing the previous night. He tipped up a laundry basket to reveal half a dozen virtually identical soiled white shirts, declaring that he had no idea which he'd worn the previous evening. None showed any sign of bloodstains.

Stares hardened when de Marigny produced the brown jacket and slacks he had worn to the party. "This suit is freshly pressed," said Melchen. "How do you account for that?"[9]

The local commissioner of police, Colonel R. A. Erskine-Lindop, who was also on hand, came to de Marigny's defense, confirming that it was *de rigueur* in the tropics to have one's suit pressed after it had been worn. "It gives one a very good presence," he said, "and is a very British tradition."[10]

This wasn't enough to pacify Melchen and Barker. They presented their findings to the local authorities, and the next day, scarcely thirty-six hours after the discovery of Oakes's body, de Marigny found himself behind bars, charged with murder and facing the gallows.

The Notorious Fingerprint

The crucial evidence—the evidence that the Crown believed would hang de Marigny—was a single fingerprint found in Oakes's bedroom. Although the individuality of fingerprints had been suspected, if not proved, in biblical times, it wasn't until the late nineteenth century that a system of identifying and codifying fingerprints came about. It remains the greatest advance yet seen in crime detection. Even DNA typing, as miraculous as that might seem, cannot offer the same degree of absolute certainty, because it cannot differentiate between identical twins. Say one twin decides to commit a crime; any DNA left at the scene would be valueless to the prosecution, if the defense could demonstrate that either twin had the opportunity to commit the crime. But if the crooked twin happens to touch something, the game is up. The DNA might be identical, but the fingerprints never match. In more than a century of use, in every country on earth, no two people have ever been shown to share the same fingerprint.

Melchen was certain he had the goods on de Marigny. The way he told it, when he and Barker arrived at Westbourne just after lunchtime on the day of the murder, they had searched for fingerprints in Oakes's bedroom. In the top section of the Chinese screen they had found a clear print that was later identified as belonging to de Marigny. Since de Marigny hadn't reached Westbourne until three or four o'clock that afternoon—as attested to by two Bahamian policemen—the inference was inescapable: De Marigny had obviously left the print during the commission of the crime.

When this news reached de Marigny, penned up in a cockroach-infested cell at Nassau jail, his spirits soared. Melchen was wrong! Not only had de Marigny been at Westbourne on the *morning* of the killing, long before the Miami detectives arrived, but he also had the best possible corroboration. "Colonel Erskine-Lindop . . . was with me," de Marigny excitedly told his counsel, Godfrey Higgs. "He can vouch I was there at that time."

"I hope he can."[11] Higgs didn't sound enthusiastic.

Just one week before the preliminary hearing, Higgs's skepticism was vindicated. Out of the blue, it was announced that Erskine-Lindop had been transferred from the Bahamas. With immediate effect he was taking up a position as assistant commissioner of police in Trinidad, fourteen hundred miles away, and would therefore be unable to testify at the trial. The timing was suspicious in the extreme. Somebody, it seemed, was prepared to go to any length to make sure de Marigny was hanged.

De Marigny harbored few doubts as to who that person might be. Ever since the finding of Oakes's body, the Duke of Windsor had overseen every aspect of the investigation. At first he had maintained control over the investigators by phone, but one day after the killing he put in a personal appearance at Westbourne. With an imperious gesture he had beckoned Barker, and the two men had disappeared upstairs. No record was kept of the thirty-minute conversation. When the duke emerged he left without saying a word. Two hours later, de Marigny was charged with murder.

Even more curiously, right before de Marigny went on trial for his life. the Duke and Duchess of Windsor took off for America. Rumors swept the island that the duke very much wanted to be incommunicado for the duration.

The sweltering little courtroom in Nassau was packed to the rafters with journalists when de Marigny faced his accusers on October 18. This was the biggest trial since the Hauptmann fiasco in 1935, and each day hundreds of thousands of words were cabled to every corner of the earth.

The Crown opened its case by painting de Marigny in the most sinister light possible. First there was motive: strapped for cash, fearful that his wife might be disinherited, he'd also been heard to physically threaten Oakes. He had opportunity: witnesses placed him near Westbourne at roughly the time of the murder. When it came to means, one police officer testified that de Marigny had asked him if, under British law, a man could be convicted of murder if the murder weapon was not found or on just circumstantial evidence alone. (The answer was "yes" to both questions.)

But no one doubted that this case would stand or fall upon that solitary fingerprint lifted from the Chinese screen.

Melchen began his testimony well, with a lengthy description of the equipment he and Barker had brought with them to Nassau: two microscopes, a Speed Graphic camera, and a fingerprint outfit containing dusting powder, a camel hair brush, Scotch tape, rubber patches of the type used to mend tire punctures, a small pair of scissors, and a magnifying glass. If Melchen was hoping to convey the impression of the well-prepared expert from Miami, he succeeded wonderfully well. But it was all downhill from there.

Erskine-Lindop might have conveniently disappeared, but Dorothy Clark had not. She had been a guest at de Marigny's party, and her testimony proved doubly useful to the defendant. Not only could she confirm that he had, indeed, scorched his arm trying to light a candle at the table, but also she was able to place him at Westbourne on the morning of July 8. She herself had been there at that time, and distinctly recalled de Marigny going upstairs at between eleven and twelve o'clock, well before the plane carrying Melchen and Barker had touched down at Nassau Airport.

When Melchen heard this, beads of sweat suddenly prickled his broad forehead. At the preliminary hearing, he and two Bahamian constables had stated categorically that de Marigny had not entered Oakes's bedroom until 3:30 P.M.—*after* the fingerprint had been found. Now Melchen was forced to recant. In a humiliating climb down he admitted, "It was a mistake."[12]

"What a mistake!" Ernest Callender, de Marigny's assistant counsel, dripped sarcasm. "What a coincidence that you *and* two constables should make the *same* mistake!"[13]

The more Melchen blustered, the more garbled his answers became. Sensing that the impetus was slipping away and desperate to save his own skin, he blurted out that he had not learned that the fingerprint was that of the defendant until July 15, when he had attended Oakes's funeral in Maine.

Callender stared askance. "Did you and Barker travel [to Maine] together?"

"Yes."

"Did you discuss the fingerprints?"

"No, we did not discuss the fingerprints."[14]

A murmur of disbelief rumbled around the court. It seemed barely credible that two detectives, working the biggest case of their careers, would not have mentioned the solitary piece of hard evidence against the defendant. Whether through heat, humidity, or tension, Melchen visibly wilted under the pressure and limped from the stand.

It was left to Barker to repair the damage. Confident, self-possessed, altogether a tougher proposition than his colleague, he began by saying that he had lifted so many fingerprints at the house that he had run out of Scotch tape and had to use a rubber patch for the last three, among which was the print lifted from the Chinese screen.

"Lifting" a print means just that. First it is dusted with powder to make it more visible, then physically removed from the surface, either by tape or, more commonly nowadays, rubber "lifters." In this technique a thin, transparent celluloid cover is removed from the adhesive side of the lifter, which is then pressed carefully over the developed print. The lifted print is re-covered with the celluloid for storage and transportation. In cruder form, this was the technique that Barker had employed. Higgs wanted to know why the print had not been dusted and photographed *in situ*.

"I did not have my fingerprint camera with me," said Barker, ". . . as I was proceeding on the theory that the case was a suicide and did not involve any criminal act, I saw no need to bring the latent-fingerprint camera with me from Miami."

"So, by your process of lifting this print on the rubber matting, you deliberately destroyed the best evidence, which was the print itself?"

"The manner of lifting the print does destroy it, yes." [15]

When asked to point to the area on the screen where the print had been found, Barker floundered. Higgs pushed hard. "Have you ever introduced as evidence a lifted fingerprint without first having photographed the actual impression as it was found on the object?"

"Certainly, scores of times."

"Give me the names of the cases."

Caught off guard, Barker stood open-mouthed for a moment, then shook his head. "I can't." [16]

In his zeal to regain the initiative, Barker succeeded only in deepening the hole around himself. His claim that he had not dusted the bed headboard for prints because "the heat from the fire would have destroyed the latent friction ridges" [17] brought a puzzled frown from Higgs.

"Why would a latent fingerprint be preserved on the screen," Higgs asked, "which is blistered due to heat, and not on the headboard?"

"I could tell by looking for certain that there would be no fingerprints left on the bed."

"And did you not dust the footboard of the bed either?"

"There is no footboard of the bed there."

Higgs held up a photograph of Oakes's bedroom. "What, pray, is that?"

Barker remained silent for several moments. All his earlier cockiness evaporated in an instant. Finally he mumbled, "I see that the bed does have a footboard. I did not dust it." [18]

Melchen had watched this exchange with an expression of mounting horror. Unable to stomach any more, he rushed outside and was physically sick, cursing Barker to all in earshot.

Evidence of Frame-up

It had been a miserable performance from Barker and Melchen, but so far there was nothing to suggest anything other than slipshod detective work. The testimony that stood this notion on its ear came from two defense fingerprint experts, Maurice O'Neill of the New Orleans Police Department, a past president of the International Association of Identification, and Leonard Keeler. They contended that the fingerprint produced in court could not possibly have been lifted from the Chinese screen. While fascinated jury members craned their necks for a better view, they lifted another print from the screen's scrolled, lacquered surface. This showed obvious pattern traces in the background, whereas the de Marigny fingerprint was clean and clear, as if it had been lifted from a smooth surface. The screen print must have come from somewhere else.

One possible source emerged when de Marigny took the stand. He recalled a curious incident at Westbourne, one day after the killing. He had been discussing the case with Melchen, when the detective suddenly steered him toward a small table, on which were a pitcher of water and two glasses. In a casual tone he had asked de Marigny to pour a glasses of water for them both. De Marigny drained his in a single swallow, then accepted the pack of Lucky Strike cigarettes that Melchen handed him. Without thinking, de Marigny had taken one and returned the pack.

In the midst of this exchange, Barker had put his head around the door and asked, "Is everything okay?"

"Yes,"[19] Melchen replied and almost immediately brought the interview to a close.

At the time the incident had passed unnoticed. Now it assumed all kinds of ominous implications. A fingerprint lifted from either the glass or the shiny cigarette pack would have produced exactly the kind of result as the disputed print.

Throughout the trial the judge, Sir Oscar Daly, had shown himself favorably disposed toward the defendant, while becoming increasingly impatient and distrustful of the police tactics. Finally he aired his misgivings publicly, as he taxed Higgs about the print alleged to have been lifted from the screen. "What you say is . . . it might be a forgery?"

"That is exactly my fear and contention,"[20] replied Higgs.

As if by magic, a somber cloud lifted from the courtroom. After days of the heaviest hints imaginable, at last it was in the open—an unequivocal declaration that someone had attempted to frame the defendant.

Whether the jury discussed who that someone might be, we shall never know. What is certain is that on November 14, by a 9 to 3 majority, they found de Marigny not guilty. Perhaps to placate the minority on the panel who voted for conviction, they added a rider that the defendant be deported, which he duly was.

Not guilty, but not innocent enough to be accepted back into Nassau society—that was de Marigny's fate. Although Nancy had stood stoically by him during the trial—"I had no desire to be the widow of a man hanged for murder,"[21] she later remarked sourly—once the verdict was returned, she wasted little time in washing her hands of this bothersome paramour, and the marriage folded.

As for de Marigny, he continued his swashbuckling ways for some years to come, until he was eventually granted American citizenship. He died at his home in Houston on January 28, 1998.

Nowadays de Marigny is customarily depicted as the unwitting dupe in a frame-up that went sorely awry. In his autobiography he rails with justifiable bitterness against the colonial British establishment that tried so hard to hang him. Yet nowhere is there any mention of just how fortunate he was. When the trial began, few in Nassau would have given a prayer for his chances. But Barker and Melchen changed all that. Trapped in lies, exposed as cheats, the two hapless Miami detectives turned the trial on a dime. By trying to gild the forensic lily, they succeeded only in gutting the prosecution. Suddenly there was sympathy for the defendant, just enough to gain that all-important acquittal.

Why Barker and, to a lesser extent, Melchen, chose to stand up in court and perjure themselves remains a mystery. Of one thing we can be sure: this wasn't any unilateral decision. They were merely following orders. The only conclusion that makes any sense is that the authorities—and that means the Duke of Windsor—genuinely believed de Marigny was guilty, but, lacking the proof to convict him, they were prepared to bend the evidence. In requesting the Miami Police Department in 1943 to investigate the murder, the duke knew he was enlisting the services of one of America's most corrupt law enforcement agencies. Melchen was infinitely malleable, while Barker was dogged throughout his career by problems with drugs and close ties to organized crime.*

---

*Barker was shot to death by his own son on December 26, 1952. The killing was ruled "justifiable homicide."

Someone else with rumored links to the Mob was Harold Christie, and whenever talk gets around to "who really killed Sir Harry Oakes," his is usually the first name in the frame. Certainly his courtroom performance—constantly mopping the sweat from his brow, hesitant, and rambling—bore all the hallmarks of guilt, as did several of his answers. For instance, he swore under oath that neither he nor Oakes had left Westbourne after 11:00 P.M. that night; yet Captain Edward Sears, a local policeman, who had known Christie since their school days, confidently testified that he had seen him on George Street, Nassau, at 1:00 A.M. on the night of the murder. Christie had been the passenger in a station wagon speeding away from Nassau Harbor. Sears could not identify the other man in the car, the driver. He knew only that it was a white man, a stranger to the island.

All Christie could offer in reply was that Sears had been mistaken. Local gossip had Christie as a pawn of American gangster Meyer Lansky, whose dreams of establishing a casino in Nassau were being thwarted by Oakes's antigambling intransigence. Was the passenger in Christie's car an imported hit man? We'll never know.

Although de Marigny discounted tales of Mafia involvement, he always believed that Christie was behind the murder of Oakes. His reasoning is rather more arcane. In this version Oakes was planning to abandon Nassau for Mexico, where he had been illegally stashing money for years, leaving Christie high and dry on some dubious land deals. If there were any shady currency transactions, it might account for what has become the second great mystery in this case: What happened to Oakes's missing millions? When probated, his will amounted to a shade under $12 million—still a vast sum, but a tiny fraction of most estimates.

Often overlooked by commentators is one glaring anomaly in Christie's testimony. On that first tumultuous morning, babbling to a journalist, he stated explicitly that Oakes had been *shot!* (De Marigny always maintained that a Nassau doctor told him Oakes was really the victim of gunshots.) Yet the autopsy recorded the cause of death as blows to the head. Even at the height of a tropical storm, it is hard to imagine that Christie could have slept through four gunshots fired twenty or so feet away. Was this a slip of the tongue, or was it something more sinister? Conclusion or confusion?

The answer to this conundrum, say some, is to be found in Oakes's home state of Maine, where his body lies to the present day in a crypt in the East Dover Cemetery at Dover-Foxcroft. If the rumors are true, the skull might still contain four small-caliber bullets.

Of course, it might not.

Chapter 8

Samuel Sheppard (1954)

Medical Malpractice and Dr. Sam

This is a story of two Dr. Sams. The first, Sam Sheppard, holds a unique place in American legal history, being the only person to be convicted, acquitted, and then "convicted" again, all for the same murder. In the eyes of many, this Ohio physician remains a martyr to the dreadful power of runaway media, a tragic example of what can happen when some delusional newspaper editor decides to scrap the Constitution, don the mantle of Torquemada, and hound an innocent man into jail. In the other camp lurks a sizable body of opinion that believes Sheppard was a wife-murdering adulterer who got lucky with his lawyer and ultimately dodged justice. It's true to say that nobody harbored this opinion more than the other Dr. Sam.

He was Samuel R. Gerber, the Cuyahoga County coroner, a small, fussy man with a hankering for bow ties and an insatiable thirst for publicity. He had first hit the headlines in November 1936, when he was drafted in to help track down the so-called Butcher of Kingsbury Run, a serial killer who had successfully eluded Cleveland's finest detectives for more than a year. The inquiry leader, Eliot Ness, another law-enforcement hotshot with a prodigious talent for self-promotion, didn't take kindly to this newcomer's meddling, and soon both men were snarling at each other's throat. (They never did catch the Butcher.)

For almost fifty years Gerber ruled the roost as coroner, and in that time he gained a reputation for tenacity and innovation that made him the template for the TV pathologist *Quincy*. Gerber didn't just preside over inquests, he often quarterbacked whole investigations. Egocentric and arrogant, he was the kind of man who made foes easily, and from all accounts he hated the Sheppard family. This animus was rooted in the fact that the Sheppards—a father and three sons—ran an osteopathic clinic, which to an old-school medical man like Gerber was akin to hawking snake oil. With its lavish claims to cure all manner of ailments through manipulation of the musculo-skeletal system, osteopathy still arouses suspicion in many medical quarters. Back in the deeply suspicious 1950s, its practitioners were regarded as closet witch doctors. This didn't prevent osteopathy from being extremely profitable, as the Sheppards with their fine houses and imported sports cars could glowingly attest. So maybe it was just an advanced case of jealousy that reportedly drove Gerber to bad-mouth the Sheppards to an intern in June 1954, swearing, "I'm going to get them some day."[1]

He was as good as his word. More than anyone else it was Gerber, with his fantastic witness stand utterances, who convicted Sheppard when the thirty-year-old doctor was first tried for murder in 1954. This made it only fit and proper that it should be those very same inanities that provided the key to Sheppard's acquittal when he was retried in 1966. The irony is that had Gerber and the rest of his cronies eased up on the retributive gas, then there would have been no second or third trial, no American Dreyfus, and we would have been denied one of the most fascinating forensic mysteries in history.

It began like this.

In the early hours of July 4, 1954, someone battered Marilyn Sheppard, age thirty, to death at her fashionable home on the shores of Lake Erie, in Bay Village, close to Cleveland. That much is certain. What part, if any, her husband played in this savage assault is a question that has exercised the American conscience ever since. For the rest of his life, Sam Sheppard would always claim that he was downstairs dozing on the living room couch when his pregnant wife was attacked. His first hint of trouble came, he said, when he was abruptly awakened by a sudden moan or scream. Rushing upstairs, he saw "a white form"[2] leaning over Marilyn, who lay groaning on the bed. A second later, so he said, everything went black.

When he came to, he realized he had been clubbed. His first instinct was to check Marilyn's pulse. Nothing. Semi-naked and drenched in blood from the gaping head wounds that ended her life, she lay sprawled across the bed. Suddenly a noise came from downstairs. Sheppard ran down and saw an intruder—a tall man, six feet, three inches with bushy hair—by the back door. In the chase that followed, Sheppard pursued the intruder down some steps and on to the beach, where he tackled him and a brawl broke out. Weakened by the earlier blow to his neck, Sheppard proved to be no match for his burly foe and was soon choked into unconsciousness.

This time when he regained his senses, he was stripped to the waist and lying half submerged in the lapping waves of Lake Erie. With his mind reeling, he staggered back to the house and phoned for help.

Evidentially, this is where Sam Sheppard's problems began. Instead of calling the police, at 5:45 A.M. he phoned a neighbor, Spencer Houk, the local mayor. "For God's sake, Spen," he said, "get over here quick. I think they've killed Marilyn."[3] Only at 5:58 A.M., after Houk and his wife, Esther, had arrived at the Sheppard house, were the police summoned. The first officer showed up just four minutes later. Judging from the general disorder he found, this bore all the hallmarks of a robbery gone horribly wrong.

Mistake number two occurred when Sam's brother, Richard, who arrived at 6:10 A.M. and rushed upstairs to examine Marilyn, was later allegedly overheard by Houk saying: "Did you do this, Sam?"

"Hell, no!"[4] Sheppard snapped back.

Later, Richard vehemently denied having said anything of the sort, while Sam claimed he was so woozy he couldn't remember what was said at the time.

Mistake number three—and arguably the biggest blunder of all—came when the Sheppard brothers whisked Sam away by station wagon to the family clinic. They claimed he needed emergency treatment for the life-threatening injuries he had suffered. Others, especially the frustrated local newshounds, reckoned that the Sheppard boys were closing ranks.

By the time Gerber descended on the crime scene at 8:00 A.M., the Sheppards were long gone, which didn't please the tetchy coroner one little bit as he went about his work. He estimated the time of death at between 3:00 A.M. and 4:00 A.M., noting also that Marilyn's watch had stopped at 3:15 A.M. Then he did some figuring. If the watch could be trusted, two hours and a half hours had elapsed between the murder and when Sheppard first called Houk. Could he really have been unconscious for that long? Or did this hiatus have some other implication?

Already Gerber was suspicious. The downstairs crime scene just didn't ring true. Yes, some drawers had been yanked from a desk and tossed aside. Yes, the contents of Sheppard's medical bag lay strewn across the floor. But compared to the bloodbath in the bedroom this looked trivial, staged almost. Other irregularities surfaced: There was no obvious sign of a forced entry to the house, and considering all the mayhem that had occurred in the house, the Sheppard's dog had remained remarkably quiet. The subsequent discovery of a canvas bag containing Sheppard's blood-flecked wristwatch, key chain and key, and a fraternity ring, tossed into some shrubbery outside the back door, only reinforced Gerber's view that a robbery had been faked in order to conceal a domestic murder.

Within hours of Marilyn Sheppard being murdered, Gerber was convinced that her husband was the killer: except there wasn't a clue to suggest why an apparently upstanding young osteopath had suddenly gone berserk and battered his wife to death.

Then came the revelation that blew the case wide open—Sam Sheppard had been screwing around.

They were just rumors at first, rumors that Sheppard foolishly denied at a rowdy inquest presided over by Gerber. Confirmation of his perjury came in the ravishing form of Susan Hayes, a lab technician, who broke down under questioning and admitted to a long-running affair with the handsome doctor. Motive with a capital M, reckoned the police, enough for murder. The local media bayed their agreement.

Ever since the killing, Louis B. Seltzer, editor of the now defunct *Cleveland Press*, and a major player in Ohio politics, had been spearheading an anti-Sheppard witch-hunt of quite appalling savagery. He filled acres of newsprint and sold thousands of extra copies with his lurid speculations, working himself into a lather of self-righteous indignation over perceived police tardiness in not charging the obvious suspect. Eventually, goaded beyond endurance by Seltzer's ravings—"Quit Stalling—Bring Him In!" screamed the July 30 headline—the police lost their nerve. That night Sheppard was charged with murder.

Police drawing of "bushy-haired stranger," wanted for the murder of Marilyn Sheppard.

Viewed objectively, his story did contain puzzling inconsistencies. His earliest statement had him seeing "a white form" standing over his wife, then being clubbed unconscious from behind, a clear reference to two assailants. This was borne out in his phone call to Houk: ". . . *they've* killed Marilyn." In later statements, all mention of two people disappeared, with the attackers abruptly condensed to a single "bushy-haired man."

And then there was Sheppard's incongruously pristine appearance. Given the slaughterhouse state of the bedroom, it defied belief that the killer could be anything other than soaked in blood, yet all Sheppard had to show, after allegedly grappling with this intruder long enough to be choked senseless, was a single smudge of blood on his pants. Admittedly he had been doused by the chilly waves of Lake Erie—and nothing gets rid of blood quite like cold water—but there was another problem: his T-shirt was missing. By his own admission he had been wearing a white T-shirt when he fell asleep on the couch that night, yet somehow it disappeared, never to be seen again, with Sheppard unable to account for its loss.

It was an odd, even unlikely series of events. But did it amount to proof of murder?

Circus Time in Cleveland

A sober, serious-minded trial might have gotten to the bottom of this puzzle, but what the state of Ohio inflicted on Samuel Sheppard, with its popping flashbulbs and bare-knuckle inquisitorial ferocity, was right out of the Joe McCarthy circus manual, a real bear pit. Even before it began, the trial judge, Edward Blythin, confided to noted journalist Dorothy Kilgallen, "[Sheppard's] as guilty as hell. There's no question about it."[5] Once in court, Blythin let the witnesses run wild.

What Seltzer had been allowed to do in the public arena, Gerber did on the witness stand. Venomous and vindictive beyond any normal call of duty, he brazenly admitted "delving into"[6] the possibility that the defendant had been sterile, to investigate rumors that Sheppard had killed Marilyn because she was carrying another man's child. (In fairness to Gerber it should be mentioned that members of the Sheppard family had also smeared Marilyn's reputation, suggesting that she had been killed by a thwarted lover, so the trash wasn't flying all one way.) Gerber's genetic probing had even extended to questioning the paternity of Sheppard's six-year-old son, Chip, who had been asleep at the time of the murder in a bedroom just along from where Marilyn was slaughtered.

But Gerber's deadliest thrust came when he began theorizing about the murder weapon, which had never been found. In a breathtaking flight of fancy, he asserted that a bloody imprint on the pillow beneath Marilyn's head had been made by a "two-bladed surgical instrument with teeth on the end of each blade,"[7] then hinted darkly that this was the missing weapon. For an impressionable jury listening to an expert witness, the link between "surgical instrument" and "doctor defendant" was too alluring to ignore, and on December 21, 1954, Sheppard was convicted of second-degree murder and later jailed for life.

While Cleveland's newspapers whooped their delight, a small faction of out-of-town journalists was utterly dismayed by the verdict. The way they saw and reported it, Sheppard had been convicted for adultery, nothing more. Thanks largely to their well-documented outrage, Sheppard's name stayed in the public arena and led to his story being adapted as the hugely successful TV series *The Fugitive.* Each week millions sympathized with the patently innocent and marvelously angst-ridden Dr. Richard Kimble as he battled all odds to track down the elusive "one-armed man" who had really killed his wife, and there can be little doubt that for some impressionable viewers, Kimble and Sheppard became interchangeable players in the same story. As the viewing figures soared, so did concerns about the unfairness of Sam Sheppard's trial.

Finally, in 1964, a judge agreed and freed Sheppard on bail, saying, "Freedom of the press is truly one of the great freedoms we cherish; but it cannot be permitted to overshadow the rights of an individual to a fair trial."[8] The following year the U.S. Supreme Court agreed, setting the conviction aside because Judge Blythin had failed to protect Sheppard ". . . from the massive, pervasive, and prejudicial publicity that attended his prosecution."[9]

Understandably the Ohio press was less than thrilled by this legal tongue-lashing. The *Cleveland Plain Dealer*, in a rare display of unanimity with its rival, whined that in "zealously" guarding a defendant's right to a fair trial, the court "has narrowed the field open to the free press."[10]

Defiant and unbowed, the state of Ohio once again tried Sam Sheppard for murder. This time the prosecution was decidedly low-key, minus all those tantalizing sex interludes that had so spiced up the first trial. Making it even worse, the state witnesses now found themselves being grilled by one of the country's toughest defense lawyers.

F. Lee Bailey was young, hard-hitting, bulging with ambition, and he bulldozed a path right through the prosecution case. His main ammunition came in the form of a report by noted criminalist Dr. Paul Kirk, who in 1955 had reexamined the crime scene evidence and reached several interesting conclusions. Analysis of spatter marks in the bedroom, made as blood flew from the bludgeoning murder weapon, led him to conclude that the assailant was left-handed—Sheppard was right-handed—and he further stated that, in his opinion, Marilyn had bitten her attacker, because in wrenching away from the bite the killer had broken off two of Marilyn's teeth. This was crucial because, when examined immediately after the murder, Sheppard displayed no evidence of bite marks. Then there was the spread-eagled position of the body, strongly suggesting, said Kirk, that Marilyn had been raped.

These were big points for Bailey, and he kept slugging away. But he saved his Sunday punch for the strutting Samuel Gerber. Bailey wanted to know more about this mysterious "surgical instrument" imprint on the pillow. While Gerber blustered over the large bloodstain—which resembled nothing more than a crimson Rorschach blot—and stubbornly stuck by his original opinion, Bailey jeered: "Well, now, Dr. Gerber, just what kind of surgical instrument do you see here?"

"I'm not sure."

"Would it be an instrument you yourself have handled?"

"I don't know if I've handled one or not."

"Have you ever seen such an instrument in any hospital, or medical supply catalog, or anywhere else, Dr. Gerber?"

"No, not that I can remember." Realizing the weakness of his response, Gerber, in his desperation to redeem himself, blurted out that he had spent the last twelve years "[hunting] all over the United States" for just such an item. Bailey could hardly believe his ears.

"Please tell us what you found," he asked.

Too late, Gerber realized that he had trapped himself. "I didn't find one,"[11] he muttered feebly.

As in the first trial, it was "expert testimony"—or in this case its flagrant lack—that won the day. On December 16, 1966, fewer than twelve hours of jury deliberation ended Sam Sheppard's twelve-year ordeal. But it was liberty at a price. With his health in tatters, he died in 1970.

And that should have been an end to it, except that in 1995 Sam Sheppard's son, Chip, decided to file suit against the state of Ohio, claiming wrongful imprisonment of his father. Now calling himself Sam Reese Sheppard, he wanted a clear, unequivocal declaration of innocence for his father, and an admission of culpability from the state. And he also wanted a couple of million dollars.

It was time to try Dr. Samuel Sheppard for the third time.

Applying the latest forensic techniques to a forty-five-year-old murder mystery was never going to be easy. Much of the evidence had degraded, much had gone missing. Original witnesses, too, were getting scarce: Gerber had retired in December 1986 and died five months later; both Houks were also dead. Still, there was plenty to savor when the combatants eventually squared off in a Cleveland courtroom on February 14, 2000.

Because this was a civil action, the eight-member jury would reach their decision on the "preponderance of the evidence," rather than the more stringent requirements of a criminal trial, which requires guilt to be established "beyond a reasonable doubt." On balance this favored the plaintiffs: they had to demonstrate only that Sam Sheppard had probably not killed his wife. However, as any lawyer knows, proving innocence can often be far tougher than establishing guilt. Not that the Sheppard estate was overly concerned by this requirement; they were packing what they considered to be the ultimate weapon—a brand-new suspect.

First, though, Terry Gilbert, chief lawyer for the Sheppard team, needed to update some points from the original trial. At that time the defense argued that Sheppard's injuries could not possibly have been self-inflicted and must have occurred when he grappled with the intruder. Half a century on, and Dr. William Fallon, a Cleveland trauma expert, agreed. He had reviewed Sheppard's 1954 medical records from the period when he was sequestered at the family-owned Bay View Hospital. These listed three major injuries: trauma to the head, face, and neck; a concussion injury affecting the brain; and evidence of a spinal cord contusion.

Sheppard was X-rayed twice while in the hospital, the first time within hours of the murder. This film clearly showed a chip fracture to the second vertebra, strong corroboration for Sheppard's claim to have been hit on the back of the neck when he entered the bedroom. In Fallon's opinion it would have been virtually impossible for Sheppard to have inflicted such grievous injuries on himself. Also on this X ray was clear evidence of congenital joint disease, or degeneration, on two other vertebrae.

This puzzled the state prosecutor, Steve Dever, because on the second X ray, taken just forty-eight hours later, this joint disease had miraculously disappeared. Come to that, so had the chip fracture!

Thrown onto the back foot, the plaintiffs were stymied, unable—or unwilling—to explain these discrepancies.

Dever reckoned he had the answer. To support his hunch he called Dr. Robert J. White, an internationally renowned Cleveland neurosurgeon, who testified that Sheppard's injuries had been grossly exaggerated by himself and his brothers. White said that Stephen Sheppard's description of his brother's injuries, as noted on his medical chart, was unsupported by two other physicians who had treated Sheppard on July 4, 1954. As for those mysterious X rays, White could only conclude that the first film had been "substituted."[12]

Illegal Legerdemain

That's right, said, Dever, a switch had taken place, engineered by the Sheppard clan. Aware that Sam was a murder suspect and willing to do anything to save him, the family had conspired to exaggerate his injuries and substituted someone else's X ray to create a false report. Why else, if Sheppard's injuries had been so life-threatening, had he been transported to the hospital in his brother's station wagon, even though an ambulance was already at the scene? And why was Stephen Sheppard so anxious to erect a medical firewall around his brother, knowing that a major crime investigation was under way, when subsequent examination of Sheppard's injuries revealed them to be far less severe than originally thought?

The more Dever honed his conspiracy theory, the more plausible it sounded.

Desperate to regain the initiative, the Sheppard estate quickly moved on to a discussion of whether there was evidence to suggest the presence of a prowler on the fateful night. According to forensic psychiatrist Dr. Emanuel Tanay, the brutality of Marilyn Sheppard's murder pointed to a psychopathic intruder and not her husband. He said that the killing, which he termed a "classic sexual sadistic homicide,"[13] was committed by an antisocial individual who derived a great deal of pleasure from Mrs. Sheppard's pain. He told the jury that this "mutilation of a helpless pregnant woman" was not a "typical"[14] spousal homicide, and in forty years he'd never seen a domestic homicide of this nature.

Not so, said Gregg O. McCrary, a former FBI agent who had studied hundreds of crime scenes in his work for the FBI's National Center for the Analysis of Violent Crimes, and who was the author of a crime-classification manual. The extraordinary ferocity of the head injuries sustained by Marilyn Sheppard inclined him away from the archetypal sadist, who extracts his pleasure from protracted torture. This attack had been brief and explosive, far more in keeping with domestic homicides. McCrary also dashed cold water on earlier testimony that because the Sheppard marriage had no record of spousal abuse, Sam should be disregarded as a suspect, saying that only 50 percent of victims of domestic homicide had suffered previous physical abuse.

Drawing up a template of how people are "supposed" to kill is like walking into a psychological minefield. Any attempt at absolutism can blow up in one's face. Earlier in the trial, state attorney William D. Mason had pounded his fist into his hand twenty-seven times, the number of blows Marilyn Sheppard suffered, then asked the jury, "What kind of a man would go into a woman's bedroom in the middle of the night and strike her violently twenty-seven times? Is it a burglar or is it an enraged husband?"[15]

The short answer is: nobody knows.

Despite what some psychomedical practitioners would have us believe, there is no surefire method of predicting how any one person will react in any given situation. Husbands have murdered wives with a single blow; intruders have battered complete strangers to an unrecognizable pulp. Statistical analysis of violent assault may suggest a probability factor, but blinkered adherence to the principle of detection by percentages is as dangerous as it is unproven. Elsewhere in this book we see how Rachel Nickell was butchered by a stranger who stabbed her forty-nine times. On that occasion a psychological profiler suggested it was the work of a frenzied pervert. How would prosecutor Mason have explained that anomaly?

Perhaps this is why other Sheppard estate witnesses took care to distance themselves from Tanay's view. As far as pathologist Dr. Cyril Wecht was concerned, the murder was "overkill"[16] without any hint of torture, and he agreed with the state theory that this was a "rage killing"[17] possibly staged to look like a sexual assault.

Wecht, the Pittsburgh coroner and himself no stranger to controversy—he acted as a consultant on the controversial Oliver Stone movie *JFK* and is on record as saying he believes Sirhan Sirhan was not responsible for Robert F. Kennedy's assassination—had been called primarily to critique Gerber's role in the original investigation, and he didn't disappoint. He laced into Gerber, accusing him of engaging in a "bizarre" and "unacceptable"[18] rush to condemn Sheppard. This snap judgment, said Wecht, was typical of Gerber's slipshod handling of the whole case. Despite having an array of qualified experts in his office, Gerber chose to control all aspects of the investigation regardless of his own lack of expertise or qualification.

It was this arrogant self-reliance that made Gerber such a sitting duck for those forensic snipers who came after, particularly Paul Kirk. In 1966 it had been his forensic reassessment of the crime scene that had played a key role in securing Sheppard's acquittal, especially his claim that Marilyn had broken off two teeth biting her attacker, thus exonerating Sheppard, who showed no sign of any such injuries.

Now it was Kirk's turn to come under the gun. Forensic odontology has come a long way since the mid-1950s, and in asking Dr. Lowell J. Levine to reexamine Marilyn Sheppard's teeth, the state of Ohio knew it was getting one of the premier experts in this field. When Marilyn's body was exhumed in October 1999, Levine could see straightaway that she did not break her teeth while fighting off her attacker. "It just didn't happen,"[19] he told the court. The dental injuries were caused by massive blows to the jaw and face, and while this didn't eliminate the possibility that Marilyn may have bitten her attacker, it laid waste to Kirk's categorical assertion that she definitely had.

Even the plaintiffs were forced to concede this point. But after Dr. Michael Sobel, of the University of Pittsburgh, had endorsed Levine's testimony, he then dived headlong into the perilous waters of crime scene speculation. The way he saw it, Marilyn, on her back being beaten by her attacker, had tried to grab his arm, scratching his wrist and tearing her own nail. Sobel's hypothesis was axiomatic to the Sheppard estate's suit, because at the heart of their case was a photograph of a man who, they said, bore just such a scar on his wrist.

New Suspect

His name was Richard Eberling. Without this former Bay Village handyman there would have been no suit, no third trial, no potential megapayoff, for it was the Sheppard estate's contention that it was Eberling who had carried out the murder. Furthermore, they reckoned they had the evidence to prove it.

Eberling had a checkered history—part artisan, part style guru, fulltime con man. In 1954 he had run a window washing business in Bay Village and was a regular caller at the Sheppard house. Although he later claimed to have been interviewed and cleared by the police at the time of the murder, there is no evidence to confirm this. What we do know about

Mug shot of alleged killer Richard Eberling, taken four years after Marilyn Sheppard was murdered.

Eberling is that he was no ordinary handyman. He had an extravagant talent for mythologizing that proved useful in his chosen career of fraudster and occasional burglar, and helped secure a contract in the 1970s to renovate the Cleveland mayor's suite at City Hall. By 1984 Eberling's fortunes were on the climb, and he was ensconced at the home of a wealthy Cleveland widow, ninety-year-old Ethel May Durkin. When Mrs. Durkin accidentally fell at home and died six weeks later in the hospital, it was found that she had left the bulk of her $1.5 million estate to Eberling, who promptly moved to Tennessee, where he set up home with his male partner, O. B. Henderson. Five years later, following a tip-off, evidence of will forgery, and an exhumation, Eberling and Henderson were sentenced to life for murdering Mrs. Durkin. Eberling died in prison in 1998.

These are the brief biographical details of the man who, the Sheppard estate claimed, raped and battered Marilyn Sheppard to death. Underpinning that claim was testimony from Kathleen Collins Dyal, an ex-nurse at the Durkin household. In 1996 she had abruptly come into public focus with claims that, while drunk one night in 1983, Eberling had confessed to murdering Marilyn Sheppard. Dyal repeated her story on the stand: "[Eberling] said that he had killed her and that he hit her husband on the head with a pail." She also said Eberling told her that Marilyn Sheppard "bit the hell out of him" and that "somebody else paid the bill"[20] for the killing.

What should have been a forensic firecracker fizzled out badly on cross-examination, when Dyal admitted that shortly after this alleged conversation Eberling had fired her, leaving skeptics to ponder two questions:

1. What was the likelihood of a killer confessing to someone and later firing that same person, with all the attendant risks to himself?

2. What part did revenge or possible monetary gain play in Dyal's story?

Certainly Dyal's ex-husband, Dale Andrews, had no doubt. On the stand he hooted with laughter at suggestions that Dyal would have kept such a secret hidden in her breast for thirteen years, until 1996, which was, coincidentally, right around the time that the multimillion-dollar Sheppard estate lawsuit began hitting the headlines.

Having failed dismally with their attempts to inculpate Eberling through alleged "confessions," the Sheppard estate realized they needed to fight back—fast.

For years they had been drip-feeding stories to the media about how DNA analysis had "cleared" Sam Sheppard of murder; now they had the chance to prove it. Their hopes were pinned mainly on two crime scene bloodstains. The first had been found on a wood chip on the basement stairs. The second came from Sheppard's pants. However, when Dr. Mohammad Tahir, an internationally renowned DNA expert hired by the Sheppard team, took the stand, he proved to be far more circumspect than his employers had been during the pretrial buildup. He frankly admitted that the samples were too degraded to permit the most refined test that can identify the exact DNA of individuals, and he had turned, instead, to PCR analysis. Polymerase chain reaction, or PCR, is a technique that is used to amplify the number of copies of a specific region of DNA, to produce enough DNA to be adequately tested.

Using PCR, Tahir was able to conclude that both stains contained DNA from someone other than Sheppard and Marilyn. It was what he said next that had so excited the Sheppard camp: the profile could not rule out Richard Eberling. Unfortunately—and largely ignored in the Sheppard team's flood of pretrial press releases—in court it became clear that Eberling was just one in literally millions of individuals who fit this same profile.

All was not yet lost for the Sheppard team. There was still the evidence from vaginal smear slides made at Marilyn's original autopsy; and this did sound promising. Once again, according to Tahir, they showed a mixed DNA profile, and while Sheppard could be ruled out in the second sample, Eberling could not be excluded from either smear. Such a finding appeared to greatly boost Sheppard estate claims that Marilyn had been raped and killed by a stranger, most likely Eberling.

But it wasn't quite that clear-cut.

Did Marilyn Have a Lover?

Tahir tossed the biggest damp squib yet with his opinion that the slides showed Marilyn had not engaged in very recent intercourse. This was catastrophic for the estate. Two different sperm types, none recent, would play right into the hands of the state and their claims that Marilyn may indeed have been engaged in a clandestine sexual relationship that drove her husband to murder. Gilbert struggled to recover the situation. But all he could salvage from Tahir was a tepid admission that Marilyn might possibly have been raped during the murder by someone who either had a low sperm count or did not ejaculate.

Stunned by this response, Gilbert appeared to panic, arguing that the age of the bloodstains made it impossible for them to be accurately typed, but it was already apparent that the much-vaunted edifice of "DNA certainty" that the Sheppard estate had so assiduously erected was beginning to show some ominous cracks.

These widened further still when the state produced the current Cuyahoga County coroner, Elizabeth Balraj. Like her predecessor in the job, Balraj spread her testimony over a wide range of specialties. Dealing first with the blood, she utterly refuted Gilbert's claim that old stains could not be typed, while the fact that Eberling's type A blood was not in the type O stain found on the door, she said, excluded him from that sample.

She was equally dismissive on the subject of sexual molestation. When autopsied, the body showed no genital trauma and, like Tahir, she thought that analysis of the vaginal smears showed there had been no intercourse at about the time of the assault.

Next Balraj turned her talents to crime scene analysis. She offered her opinion that Marilyn did not die slowly and was not tortured, bound, or restrained during her slaying, directly contradicting Tanay's view that Marilyn had been victimized by a sadist who killed for pleasure, not by an enraged husband.

Balraj also rebutted testimony about a fingernail injury suffered by Marilyn. Balraj argued that the fingernail was loosened from the finger by a blunt force, not torn away as she gouged her assailant, as Sobel had claimed. And as for the photograph of Eberling's wrist—the one in which Sobel had discerned a T-shaped scar, caused, most likely, he thought, by Marilyn's fingernail—Balraj professed herself unable to see a scar of any description.

Nor could James Wentzel, a highly experienced forensic photographer for the Cuyahoga County coroner's office. He said the photos did not contain proper scale measurements, were not shown in proper perspective, and were not of suitable quality to render a reliable skinprint analysis. In short, they were worthless.

Having rebutted most of the Sheppard team's claims, now it was time for the state to load up the big guns. Much the most telling evidence against Eberling being the killer, they argued, was the fact that he actually testified at the 1966 trial. This was because, seven years beforehand, he had been charged with stealing a ring from Marilyn Sheppard's sister-in-law. At the time Eberling volunteered the information that he had cut himself at the Sheppard residence and had bled on the basement stairs, testimony he repeated at Sheppard's second trial.

Remember, this is the man with whom Sheppard is supposed to have fought for his life, yet there was not a flicker of recognition as Eberling passed Sheppard to take the stand. Admittedly, Sheppard claimed to be in a woozy state at the time of the brawl and Eberling, at six feet and balding, didn't bear much resemblance to Sheppard's original description of the intruder—six-three with bushy hair—but it stretches credulity to the breaking point, given Eberling's testimony, to imagine that Sheppard would not have considered the possibility that this was his adversary on the shores of Lake Erie, if, in fact, there was any glimmer of a likelihood.

Certainly, F. Lee Bailey thought the idea preposterous. Questioned here at the third trial, he said, "I did not or at any time since form a belief that Richard Eberling killed Marilyn Sheppard."[21]

In 1954 Dr. Roger W. Marsters had been a young scientist who had analyzed trace evidence found at the crime scene. Now age eighty-one, he was back in court, to discuss the origin of those flecks of blood found on Sheppard's watch, which had been recovered from the medical bag.

Although conceding that he was not a blood-spatter expert, Marsters recalled that the watch, which Sheppard said must have been wrenched from his wrist by the assailant, was stained with clearly formed blood droplets. This was hugely significant. Had it been yanked off during a struggle, any blood adhering to it might be expected to smear; but droplets of the kind that Marsters says he saw, were far more likely the products of flying blood, such as would have sprayed from the dreadful wounds on Marilyn Sheppard's head during the frenzied attack. Put another way, either this watch was on the wrist of the killer, or it lay close by, perhaps on a bedside table.

If true, this raises serious questions about Sam Sheppard's story. But does it necessarily make him the killer? Or were there other suspects?

At the second trial, Bailey had sown seeds of doubt in the jury's mind by tarring Marilyn Sheppard with the stigma of adultery, and he named names—Spencer Houk, no less, neighbor and local mayor. Bailey theorized that Esther Houk had murdered Marilyn in a jealous rage after learning that her husband had been sleeping with Marilyn. It's an intriguing hypothesis, though quite why Sheppard would have connived at the subsequent conspiracy, given the level of antidefendant hostility that Spencer Houk displayed on the witness stand, remains a mystery.

Under questioning from Dever, Bailey reiterated his belief that the Houks, both now deceased, remained the strongest suspects. "I bought [that theory] initially and it got stronger over the years even until today."[22]

At the first trial Sheppard had done himself no favors by taking the stand. As the transcript reveals, he was blessed with a healthy regard for his own importance, never once missing an opportunity to remind the jury that he was a "doctor," and after a while it must have grated with some of the jurors.

The second time around, Bailey completely wrong-footed the prosecution with his controversial decision to keep his client off the stand. Exposed as a playboy liar in his first trial, Sheppard wisely basked in immunized silence as Bailey raised just enough doubts in the jury's mind to gain an acquittal.

But at this third trial, it was Hamlet without the prince. And suddenly there was a brand-new suspect to contend with. By offering Richard Eberling as an alternative killer, the Sheppard estate had believed they were strengthening their hand, beefing up their chances of an unequivocal declaration of "innocent." But what might have looked good on paper backfired horribly in the courtroom. Instead of being a straight call—was Sam Sheppard innocent?—the trial degenerated into a legal tug-of-war: Who, on the evidence presented, was the more likely killer, Eberling or Sheppard?

Few doubted which way the jury would jump. On April 12, 2000, they found for the state of Ohio. In effect they were saying that Sam Sheppard was "not innocent" of his wife's murder. It remains an appropriately ambiguous verdict for this most baffling of crimes, for despite all the searing doubts and suspicion, there wasn't a scintilla of hard evidence in 1954 to prove that Sam Sheppard killed his wife. There still isn't.

Chapter 9

Steven Truscott (1959)

A Time for Dying

In late 1966 a teeth-gritted feud that had simmered for years in the mortuaries and courtrooms of Britain finally erupted into open warfare. The flash point came when two London-based pathologists were sent advance copies of a transatlantic best seller scheduled for upcoming British release. The intention was for each to read the book, ringingly endorse its controversial subject matter, add their weighty imprimatur, and thereby provide a whopping boost to sales. It didn't work out that way. Rather, it demonstrated one of the enduring conundrums of forensic science: how similarly qualified experts, of broadly comparable experience, can analyze the same data and arrive at diametrically opposed conclusions.

The book, titled *The Trial of Steven Truscott,** had caused a sensation in Canada, with its harrowing tale of a fourteen-year-old Ontario schoolboy convicted of murder in 1959 and sentenced to hang, though this was later commuted to life imprisonment. According to the book's author, Isabel LeBourdais, Truscott had suffered a gross miscarriage of justice, an opinion shared by her British publishers, which was why they had canvassed the support of the nation's two premier pathologists, Professors Francis Camps and Keith Simpson.

---

*Toronto: McClelland & Stewart, 1996.

Both were the sons of doctors and both were hugely experienced—between them they had performed the quite staggering total of 180,000 autopsies, with Simpson narrowly holding the numerical ascendancy—and there the resemblance ended. In temperament and methodology they were polar opposites. Camps, restless and hotheaded, was a whirlwind in the mortuary, astounding students and assistants alike as, cigarette dangling from his lips, he tore into his work with manic speed and a total disregard for hygiene. On one memorable occasion, while ripping open a rib cage, he tore a rubber glove and scratched himself. Contemptuous of the risk of septicemia, he refused to don another glove and barreled on, growling, "I used to do autopsies with bare hands!"[1]

Being paid "by the body" turned Camps into a human conveyor belt—ten minutes maximum for the average autopsy—and fattened up his bank balance marvously no bad thing for a *bon vivant* with some expensive appetites. Sadly, his exuberance occasionally led him down some dubious paths, one of which was signposted "Steven Truscott."

After skimming LeBourdais's book, and perhaps mindful of a juicy fee, Camps let it be known that he was prepared to give evidence for the convicted youth in the event of a retrial. At this point it should be mentioned that for all his faults—and he had plenty—Camps was a tireless fighter on behalf of the underdog, never afraid to tilt at official windmills. Unpopular causes were meat and drink to this pugnacious Londoner, as was ruffling feathers.

Someone who could attest to this better than most was Professor Keith Simpson. Like Camps, Simpson had risen to prominence in World War II, and, like Camps, he had at first struggled to escape the shadow of the legendary Sir Bernard Spilsbury. When Spilsbury committed suicide in 1947, the race was on to fill the great man's shoes. In the early days Camps and Simpson were close friends, but gradually their relationship soured as Camps grew resentful of the way in which the headline-making cases tended to gravitate in Simpson's direction. This was no accident. The police and the courts preferred Simpson's cool, urbane witness stand manner, with no histrionics and, above all, no surprises. Camps, on the other hand, was flashy and mercurial and thought to be untrustworthy, and his status underwent a slow, painful erosion. Embittered and marginalized, he lashed out at every opportunity against the rival whom he now considered to be his mortal enemy. Then came the case of Steven Truscott, and with it a priceless chance to settle Simpson's hash, once and for all.

On the night of June 9, 1959, Lynne Harper, the twelve-year-old daughter of a Royal Canadian Air Force (RCAF) officer, failed to return to her home on the Clinton air base in southwestern Ontario. At first light next morning, Lesley Harper, frantic with worry, went house to house on the base, desperate for news of his missing daughter. He soon got a lead. Someone reported seeing Lynne just outside the local schoolyard the previous evening at about 7:10 P.M., perched on the crossbar of her school friend Steven Truscott's green racing bicycle, as the husky fourteen-year-old pedaled toward County Road, which ran north of the air base.

Harper rushed to the Truscott residence. Young Steven was the personification of helpfulness. Yes, he'd given Lynne a ride, because she'd wanted to see some ponies at a house on Highway 8. Being pressed for time, he had dropped her off at the junction of County Road and Highway 8. No problem, said Lynne, she would hitch the rest of the way. Steven explained that, as he cycled back toward the base, he had stopped on a small bridge overlooking a popular swimming hole and glanced back, just in time to see Lynne climbing into a car, which sped off east along Highway 8 in the direction of the ponies. Steven estimated that the time was about 7:30 P.M.

Steven Truscott, age thirteen. By the time of his trial, he looked much older.

Then he'd returned to the base and joined in a pick-up football game with some friends at approximately eight o'clock.

Later that day, Steven amplified his statement for the police, describing the car as a gray '59 Chevy, with lots of chrome and a yellow license plate. The officers stirred. This sounded promising. Each year thousands of tourists from Michigan, just an hour's drive to the southwest, flooded across the border into Ontario, and you could always spot those bright yellow Michigan plates. An immediate all-points bulletin was issued for a gray Chevrolet with Michigan plates, but it drew a blank.

What happened the next afternoon was highly significant. After no fewer than five interviews with Steven, investigators ordered a 250-man search team to fan out, not along Highway 8, where Steven claimed he had last seen Lynne, but on County Road itself.

Mostly the road cut through open wheat fields, but about three-quarters of a mile north of the air base, on the eastern side of the road, lay a small wooded area known as Lawson's Bush. Trooping down a tractor trail and across a barbed-wire fence, the searchers entered a tangle of elm, ash, and maple. At approximately 2:00 P.M., in a scrub-filled hollow fewer than a hundred yards from the paved road, the search for Lynne Harper came to an end.

She was lying on her back, naked except for her underskirt and blouse, which had been torn and knotted tightly around her neck. Her shoes and socks were laid out neatly close by, together with her hairband and turquoise shorts. Curiously, the zipper of the shorts was closed. Thirty feet away lay the final item of clothing, a pair of panties. The killer had tossed a few branches across the corpse in a desultory attempt at concealment. Also found nearby were bicycle tire tracks.

That afternoon Dr. John Penistan, Ontario's chief medical examiner, studied the body *in situ*. With temperatures hovering around 90°F for the past several days, decomposition was well advanced, but his first impression was that Lynne had been strangled, and he suspected that rape had at least been attempted, if not completed. Together with Dr. David Hall Brooks, chief medical officer at the air base, he spotted two small mounds of earth in the ground between Lynne's ankles, formed in all likelihood by the rapist's shoes as he struggled to accomplish intercourse. Peering closer at the two mounds, they detected a distinct pattern in the soft soil, such as might be made by a crepe-soled shoe.

After Penistan concluded his preliminary examination, the body was turned on its left side so that the ground beneath could be photographed, a simple maneuver that many years later would have crucial repercussions.

For now, though, Penistan was more engrossed in scouring the leaves and soil for traces of seminal fluid or blood. Satisfied there was none, he ordered that the body be removed to the undertaker's mortuary at Stratford, where he would examine it that evening.

Few autopsies in criminal history have so aroused such controversy, and there is no denying that it was conducted in less than ideal circumstances, a gloomy room illuminated by a single electric bulb, in a building swarming with doctors, police officers, and official photographers. Having said that, Penistan's conclusions would later withstand scrutiny from the best the forensic medical world had to offer.

After first confirming the cause of death as asphyxia due to ligature strangling, he found evidence—seminal fluid in the vagina—that sexual intercourse had occurred. Pronounced maggot activity made it impossible to tell if Lynne had been a virgin prior to this assault. Judging from the deep twig and undergrowth impressions on her body, she had been pressed hard onto the ground, strongly suggesting that she had been sexually assaulted and murdered where she lay.

But *when* had she died? The most useful of all guides—the fall in body temperature—was no longer available, since the body loses all natural heat in eighteen to twenty-four hours, and Lynne had obviously been dead longer than that. Nor was rigor mortis any help. Ordinarily, rigor begins within a few hours of death, reaches completion after twenty-four hours, then gradually dissipates until it is entirely gone after forty-eight hours. Because there are so many variables—musculature of the body, general health, and ambient temperature—it has led one pathologist, Professor Bernard Knight, to note: "If a body does not have any perceptible rigor it has either been dead less than six hours or more than forty-eight."[2] Beyond that, he says, everything is pure speculation. In this case, having noted that rigor had "almost passed off,"[3] Penistan went looking for other indicators of the approximate time of death.

The answer lay in the stomach. According to her mother, at 5:30 P.M. on the last day of her life, Lynne, eager to be out in the summer sunshine, had bolted down a meal of turkey, cranberry sauce, peas, and potatoes, finishing with a slice of upside-down pineapple cake for dessert, before leaving the house no later than 5:45 P.M. Easily recognizable remains of the meal were still in Lynne's stomach.

By and large, digestion is a fairly predictable process. According to former New York City Medical Examiner Dr. Michael Baden, "Very little interferes with the law of the digestive process. It is not precise to the minute (no biological process is), but within a narrow range of time it is very reliable. Within two hours of eating, ninety-five percent of food has moved out of the stomach and into the small intestine. It is as elemental as rigor mortis. The process stops at death."[4] However, like rigor mortis, any number of variables can affect this process. Fatty food takes longer to digest, as does a large meal; mood, too, can play its part; but nothing is more likely to disrupt the digestive process more than sudden trauma, either physical or psychological. Extreme terror can play havoc with the body's metabolism. Had Lynne's abductor tortured her over a prolonged period of time—several hours, say—it was possible, likely even, that her digestive process would have been grossly retarded.

But Penistan felt certain that Lynne, an otherwise healthy young girl, had been killed quickly. Allowing for the considerable volume and only partial digestion of the food in her stomach, he wrote in his report: "I find it difficult to believe that this food could have been in the stomach for as long as two hours, unless some complicating factor was present, of which I have no information. If the last meal was finished at 5:45 P.M., I would therefore conclude that death occurred prior to 7:45 P.M."[5]

If Penistan was correct and death had occurred no later than 7:45 P.M., and if Lynne Harper had died where she was found, then Steven Truscott has some serious explaining to do. After all, what was the likelihood that Lynne had climbed into a stranger's car and been spirited off down Highway 8, only for the driver to then turn around and drive back to Lawson's Bush, and there carry out murder—all before 7:45 P.M.?

Even before this, police officers had been deeply suspicious of Steven's manner and particularly his claim that he saw Lynne being picked up by a '59 Chevy on Highway 8, while he was stood on the bridge, almost a quarter of a mile away. When investigators checked the bridge themselves, they doubted that anyone could identify a car at that distance, to the point of being able to pick out the color of its license plates.

Steven Truscott's story was beginning to look threadbare.

Suspicious Injuries

The next day Steven was examined by Dr. J. A. Addison. Besides minor scratches on the boy's torso, Addison found, on each side of Steven's penis, "a brush burn of two or three days duration the size of a 25-cent piece."[6]

It was then exactly three days since Lynne's disappearance.

Addison's findings were confirmed by Dr. Brooks. Both felt the lesions were consistent with forcible intercourse (though neither was experienced with rape cases or rape-related injuries). Also, there was a wound on the back of Steven's leg, again two or three days old, prompting speculation that it might have been caused by the barbed-wire fence that surrounded the area where Lynne's body had been found.

In the cellar of Steven's home, searchers found the red jeans he had been wearing on the night in question, freshly laundered and hanging alone on a line. Despite being washed, faint traces of grass stains were discernible on the knees; and on one leg there was a tear that corresponded with the injury on Steven's leg. Detectives also searched for a pair of crepe-soled shoes that Steven was known to possess and was thought to have been wearing the night Lynne died. According to Professor Simpson, who wrote extensively about this case, the officers could not find them anywhere, and they were never seen again.

What the police did find was Steven's bicycle, and the tire tread pattern was identical to that found at the crime scene.

The circumstantial evidence was now piling up like a snowdrift at Steven Truscott's door. So far as the police were concerned, there had been no gray Chevrolet, no yellow license plates, no stranger stopping to give Lynne Harper a ride. It had all been the cunning fabrication of a brutal killer desperate to cover his tracks. The way they figured it, Steven had lured Lynne into Lawson's Bush with the intention of having intercourse; then, when she resisted, he had raped and strangled her. Disregard the suspect's age, they reasoned, and this case was just one more squalid sex murder, the kind of tragedy repeated countless times across Canada every year.

That night Steven Truscott was charged with murder.

Even though the police were convinced they had their killer, they still didn't have a watertight case. Far from it. When they attempted to re-create the final few minutes of Lynne Harper's brief life, gaping holes began to appear in the chronology, primarily because all the major eyewitnesses to the tragic events of June 9 were young children, and the stories they told were so contradictory, so perverse, as to confound reasoned analysis.

Ever since noon on that scaldingly hot day, kids from the base had either cycled or walked up County Road to cool off in the swimming hole, which was situated on the Bayfield River, about halfway between Lawson's Bush and Highway 8. Two youngsters at the swimming hole that evening, Gordon Logan and Douglas Oats, claimed to have seen Steven and Lynne heading north across the bridge at about 7:30 P.M. Five minutes later, said Gordon, he saw Steven return alone, cycling toward the air base. This was powerful evidence in Steven's favor. But three other children told radically different tales.

At about the time Steven and Lynne were seen cycling away from the base in the direction of County Road, one mile to the north two boys left the swimming hole and headed south on the same road. One, Richard Gellatly, age twelve, rode a bicycle; the other, ten-year-old Philip Burns, was on foot. Because the only intersection on County Road between the swimming hole and the air base was a railroad line, any road users heading in opposite directions at that time would be bound to pass. And so it proved. At about 7:25 P.M. Richard met and passed Steven and Lynne— still heading north—alongside Lawson's Bush. Thinking no more of it, Richard cycled on.

Five minutes later, it was Philip's turn to reach Lawson's Bush. He told the police that the road was empty, with no sign of Steven or Lynne. If true, this meant that shortly after meeting Richard, Steven and Lynne had slipped off the road—into Lawson's Bush.

One person Philip Burns *did* meet that night on County Road was thirteen-year-old Jocelyne Goddette. He was alongside Lawson's Bush when she came breathlessly cycling toward him, asking if he had seen Steven Truscott. When he said no, Jocelyne frowned with disappointment. As she later told the police, Steven had phoned that evening just before six and asked if she wanted to see some calves that evening, but a late supper had forced her to say no.

As she cycled slowly past Lawson's Bush, Jocelyne shouted out Steven's name several times. There was no answer. Nor was he at the swimming hole when she reached there. At about 7:45 P.M., disappointed and disconsolate, Jocelyne abandoned her search and headed home. Just minutes later she cycled past Lawson's Bush for the final time that evening and, if Dr. Penistan was correct, right past the body of Lynne Harper.

Jocelyne's statement hardened police suspicions that the sexual assault, if not the murder, of Lynne Harper had been premeditated. They figured that Steven had schemed to lure a girl—any girl—to Lawson's Bush, with the intention of having intercourse. After Jocelyne had declined to meet him, he'd gone hunting for another victim and found Lynne Harper. Once in Lawson's Bush, she had tried to fight off his advances, but the muscular fourteen-year-old, hormones throbbing in a sexual frenzy, had strangled her after gratifying his desires.

Even after Steven's arrest, new witnesses continued to come forward. Allan Oats, older brother of Douglas, claimed that he, too, had been on County Road that night, close to Lawson's Bush, and had spotted Steven standing alone on the bridge. Again, this was strong corroboration for Steven's story.

But all of these sightings paled, reckoned the police, when set against the evidence of Arnold "Butchy" George. Although this thirteen-year-old's

Lynne Harper's body as it was found in Lawson's Bush.

various statements were patchy and inconsistent, the final version he settled upon painted Steven in the direst light imaginable.

During Steven's first police interview—before the body had been found—he had mentioned seeing both Arnold George and Douglas Oats near the river at the crucial time. Subsequently, Arnold had confirmed that Steven passed over the bridge alone; however, after numerous alterations and amendments, he ruefully admitted that he had lied to help a friend. What really happened, he said, was that, after the body was discovered, Steven had asked him to provide a false alibi.

Other discrepancies soon appeared. On the night of Lynne's disappearance, so Arnold now claimed, he had been at the school playing field when Steven returned at 8:00 P.M., and already rumors were buzzing amongst the children that something had happened at Lawson's Bush. One of the boys had taunted Steven, shouting, "What did you do with Harper, throw her to the fishes?"[7] Stevens calmly responded that he had dropped Lynne off at Highway 8, nothing more.

If true—and it should be viewed with the utmost circumspection—Arnold's story was potentially devastating.

On the morning of September 16, 1959, Steven Truscott was led into the Goderich courthouse to stand trial for murder. The young defendant, who looked much older than his years, seemed in no way overawed by the proceedings, despite knowing that he was to be tried as an adult and therefore liable to the death penalty if convicted. Some attributed his stoicism to callous indifference; others called it plucky resilience. Whatever the reason, for someone so young, Steven bore up under the strain with remarkable fortitude.

With his parents looking on, he listened intently as the prosecution outlined its mainly circumstantial case: the missing shoes; the ripped jeans with grass-stained knees; the cut on his leg; the abrasions on his penis; the fact that he was the last person seen with Lynne; the tire tracks. To link him to the time of Lynne's death, a hermetically sealed jar containing the contents of the dead girl's stomach was brought into court, and Dr. Penistan described the digestive process. He repeated his belief that Lynne had met her death prior to 7:45 P.M. on the night she had disappeared. The defense mounted only a lukewarm attack on Penistan's findings; the real bombardment would come years later.

A succession of bewildered children took the stand and told such widely divergent stories that the judge, when summing up, was forced reluctantly to raise the possibility of perjury.

If ever a case cried out for illumination from the defendant, this was it, yet Steven Truscott never uttered a word. Throughout the two-week trial he maintained a sphinxlike detachment, watching each witness closely, heeding their testimony, never once letting down his guard. Exercising his legal right to silence was all well and good, but given the powerful circumstantial case against him, it is difficult to understand how—if he were innocent—any more harm could have accrued had he taken the stand. Mentally, he was equipped for the task; intelligence and aptitude tests showed him to be a "perfectly ordinary fourteen-year-old boy."[8]

In the end the verdict came quickly. On September 30, after three hours' deliberation, the jury found him guilty, and Steven Truscott was sentenced to be hanged. His parents—their deathly white faces set in expressions of utter disbelief—sat frozen as their young son was led from public view to some cell, there to await the hangman's rope.

It is nowadays a little-known fact that few in Canada at the time even knew the Truscott trial had even taken place. Because of the defendant's age, the judge ordered a total ban on media coverage until the verdict was delivered; so one can well imagine the sense of outrage when Canadians opened up their newspapers on October 1, 1959, and read that a fourteen-year-old Ontario schoolboy had been sentenced to hang. Though no one believed for a moment that the sentence would ever be carried out, six weeks elapsed before it was commuted to life imprisonment.

Overnight, public interest waned, and soon the name of Steven Truscott, convicted murderer/rapist, faded from memory.

For several years the case slumbered, only to be jolted back into life in 1966 when a journalist named Isabel LeBourdais published *The Trial of Steven Truscott,* a well-crafted and hopelessly biased account that portrayed the Canadian schoolboy as a modern-day martyr in short pants. Anyone coming to the Truscott case with no prior knowledge of the facts—which because of the press embargo meant virtually everyone—and reading just this book could only conclude that a gross miscarriage of justice had occurred. LeBourdais flayed everyone: the judge; the prosecution; the defense counsel; Canadian trial procedure; even the jury—"their educational level was not high"[9]—but most of her venom was reserved for the person she blamed for instigating this outrage, Dr. John Penistan.

Informed only by some out-of-date medical textbooks, she accused Penistan of grossly miscalculating the time of Lynne Harper's death. According to LeBourdais, the contents of Lynne Harper's stomach proved nothing, because digestion was too erratic to measure; the poor girl could have survived several hours after eating her final meal, making Steven Truscott's story of the stranger in the gray Chevrolet not only possible but also highly likely.

LeBourdais cited other reasons for believing in Steven's innocence. First, the time factor: according to LeBourdais—who posited her theory on the testimony of the three defense child witnesses—Steven had a mere five minutes in which to accomplish both rape and strangulation. Next there was the ligature—boldly, and as it later transpired, foolishly, she stated that it was physically impossible for Lynne to have been strangled with her own blouse. Then came some hoopla over a piece of the blouse, a few square inches, that were missing and to which she attached great evidential significance. Finally, she pointed to the various scratches and abrasions on Lynne's body as proof that she had been killed elsewhere, then dumped in Lawson's Bush.

These were all legitimate arguments, well worth considering, until outraged indignation got the better of reason, and LeBourdais lurched into the black hole of outright inaccuracy.

Why, she thundered, had the vaginal swabs, with their clear traces of semen, not been grouped to see if the killer was a "secretor"? Since 1925 it has been known that 80 percent of the population secrete their blood type in saliva, semen, and other bodily fluids. Surely Steven should have been tested to find if he conformed with the bulk of the population?

In fact, Steven *was* tested, he *was* a secretor, and it *was* possible to group the semen as belonging to his blood type. But this potentially crucial piece of evidence had to be discarded when it was found that not only was Lynne Harper a secretor as well, but that she shared the same blood group as Steven Truscott.

Despite such textual errors, or perhaps because of them, LeBourdais single-handedly raised such doubts over Truscott's conviction that a judicial review was ordered. At the same time, LeBourdais heard that her book was about to be released in Britain.

The Experts Fly In

Professor Keith Simpson had read the book and came away shaking his head. To his mind, LeBourdais had indulged herself "in unfounded, biased criticisms of the Canadian police, their pathologists, and the Canadian courts."[10] In October 1966, at the request of the Canadian government, he flew to Ottawa for what amounted to the virtual retrial of Steven Truscott. Also making the long flight from London was his old adversary Professor Francis Camps.

Never has one courtroom witnessed such a galaxy of forensic talent. Alongside Penistan and Simpson for the prosecution stood Dr. Milton Helpern, from New York, a medical examiner whose career had been spent in America's deadliest city and who knew more about murder than any man alive. Rounding out the prosecution team was Dr. Samuel Gerber, the Ohio coroner who had overseen the Sam Shepherd case. Appearing for Truscott, alongside Camps, was Dr. Charles Petty, an eminent pathologist from Baltimore, and Dr. Fred A. Jaffe, of Toronto.

As always, Simpson was meticulously prepared. Three months earlier he had traveled to Toronto for a conference with the authorities, and was amazed to find that the Ontario State Crime Laboratory had preserved every lab specimen pertaining to the Truscott case, every smear, every microscopic slide—everything, in fact, except for the partially-digested food matter. After reviewing all of this material, Simpson studied the much-maligned Dr. Penistan's notes. He found them exemplary, writing, "I [do] not remember, in 30 years, having seen a more thorough or painstaking report, or any more impartial deductions."[11]

Together, Simpson and Penistan set about exploring some of LeBourdais's more extravagant claims—first, the notion that Lynne could not possibly have been strangled with her own blouse. Using an identical blouse, purchased from stock, and a compliant policewoman, they were able to demonstrate that it was perfectly possible for the blouse to have been used as a ligature.

Just as easy to explain was the missing portion of blouse that had so exercised LeBourdais. After reconstructing the strangulation, Penistan pointed to the ligature and said to Simpson, "Cut it here."[12] Simpson did as requested, and the same piece that had been mislaid from the original blouse fell away. Working in the ill-lit undertaker's mortuary, Penistan, after cutting the ligature, had been absorbed by the body. In all likelihood the missing piece of cloth had fallen unnoticed to the floor, later to be swept up by mortuary staff and discarded.

After studying the police photographs, Simpson, whose knowledge of crime scenes was encyclopedic, had little doubt that Lynne had been assaulted and murdered where she lay. He could make nothing of Le-Bourdais's claim that tears and lacerations in the skin amounted to proof that Lynne had been killed elsewhere and then dragged to where she was found. A long scratch on her left leg—LeBourdais termed it a "gash"[13]—another on the rear left shoulder, and various small cuts and abrasions on the backs of her hands and legs and torso, all clearly shown in the photographs, seemed to Simpson to be exactly the kinds of marks he would expect to find had a struggling body been pressed or held down in that type of undergrowth.

Simpson's chance to present his findings came on October 5, 1966, when the nine-man tribunal of judges, convened to determine whether Steven Truscott should be granted a retrial, began hearing evidence. Mostly the court concerned itself with medical testimony, but first Jack Parish, an independent investigator, told how, one year and a day after the murder, he had gone to the swimming hole at 7:30 P.M. and stared up at the bridge 630 feet away. Dazzled by the setting sun and glare off the water, he found it impossible to make out the identity of anyone on the bridge. "I could see silhouettes and bicycles, but I could not recognize them."[14] Parish's experiment dealt a crushing blow to the testimony of those children at the swimming hole who claimed to have seen Steven and Lynne on the bridge.

Then it was time for the forensic duel to begin. Dr. Charles Petty, called to buttress defense claims that Lynne had been abducted by a stranger, murdered elsewhere, then dumped in Lawson's Bush, pointed to police photographs and said that areas of pallor on the dead girl's cheek and left shoulder indicated that she had lain, at some time after death, in a different position from that in which she was found.

But Petty had blundered. Earlier photographs, taken when the body was lying on its back, were produced, and these clearly showed no such discoloration. What Petty had mistaken for pallor was actually caused by pressure marks on the face and shoulder, made when Penistan ordered the body to be turned on its left side so photographs could be taken of the ground beneath.

Another serious setback for the Truscott camp came when Dr. Milton Helpern took the stand. Enormously experienced, having personally performed twenty thousand autopsies, and having supervised twice that number, he, too, was convinced that death had occurred where the body was found "no more than two hours after food was ingested."[15] A large, craggy man who did not suffer fools gladly, Helpern listened with ever mounting incredulity as Truscott's counsel put it to him that Lynne had been abducted and killed elsewhere. Helpern stared askance at the questioner for a moment. "What?" he boomed. "Murdered and then brought back near to her home and put there with her shoes and things scattered around? It just doesn't make sense."[16] Hastily, counsel moved on.

When Simpson testified, he did so in his customary polished manner. Well used to the ploys of advocacy, he was ready when defense counsel archly pointed out that Simpson's own textbook* didn't even mention stomach contents as a means of determining time of death. Simpson duly acknowledged the omission, then revealed that he had referred Penistan's conclusions to a colleague, a world authority on stomach emptying, Professor John N. Hunt of Guy's Hospital Medical School. As Hunt could not attend the tribunal, Simpson received permission to read out his colleague's opinion: "The conclusions drawn by Dr. Penistan tally with those which I have reached after a close study of the literature and more than 15 years personal research in this particular field. . . . I am satisfied that Dr. Penistan was entirely correct to conclude as he did under the circumstances."[17] Simpson, too, reaffirmed his belief in the strongest possible terms that Dr. Penistan, far from being the villain depicted in certain quarters of the Canadian press, had performed his duties honorably and well.

Unfortunately, this could not be said about Professor Francis Camps. Earlier, solely on the evidence of LeBourdais's book, he had brashly trumpeted the view that Lynne Harper's death could have occurred at any time from one to ten hours after eating—plenty of time for her to have been taken away, raped, and murdered in the mysterious gray Chevrolet, then driven back to Lawson's Bush. Now, hearing the full facts of the case for the first time, he realized, far too late, that once again he'd fallen victim to his own impulsiveness. He squirmed, he wriggled, he did everything possible to get off the hook, all to no avail. Utterly crestfallen, he was forced to concede that "Dr. Penistan's conclusions were very fair."[18]

It was a humiliating climb-down, made infinitely worse when, during cross-examination, Don Scott, senior counsel for the Canadian director of public prosecutions, abruptly handed Camps a document. "Look at this letter, and tell the court if that is your signature to it."

Camps flushed. "Yes, it is. It was a private letter and—"

Forensic Medicine, 1st ed. (London: Arnold, 1947).

"Never mind about that. I am going to read it to the court."[19]

The letter, written by Camps to the British Attorney General in London, contained an unsolicited offer to give evidence for Truscott should an appeal be held. A delicious gasp rippled through the court. In those more genteel times, such blatant touting for work was virtually a capital crime in medical circles, and Camps had been caught red-handed, trying to line his own pockets. Somehow the squat Londoner managed to rein in his volcanic temper as Scott continued his rapierlike probing.

Between responses, Camps threw villainous glares across the courtroom at Simpson, convinced this humiliation had been orchestrated by his archenemy. At the conclusion of his ordeal, Camps, crimson-faced with rage, rushed from the stand and out of court, not even bothering to stop for his briefcase. Scott and Simpson exchanged sweet smiles of triumph, though neither would reveal how such deadly ammunition had found its way into the prosecutor's arsenal.

Camps's participation in the Truscott case had been an unmitigated disaster; so much so that his biographer excised all mention of this Canadian fiasco from his account of the pathologist's life.*

With so many high-octane egos dueling in the Ottawa courtroom, Truscott's customarily subdued personality was even easier to overlook. This time he did volunteer evidence in his own defense, but it was too little, too late. Shorn of his youth, Truscott was denied his strongest asset. People don't expect fourteen-year-olds to rape and kill; now it was a grown man giving testimony, not a young teenager. Hesitantly delivered, often in a whisper that few could hear, his evidence threw little light on the tragedy. All he could offer was a blanket denial of everything the prosecution witnesses had stated at his trial.

Ironically, like Camps, Truscott was made to suffer for the efforts of his pen. A letter, written to the Parole Board while he was serving his sentence, begged for a "chance to prove that one dreadful mistake does not mean I will ever make another one [sic]."[20] Asked why write this if he was innocent, he replied, "To get out."[21]

To an impartial observer, this does not seem unreasonable. Parole boards are notoriously susceptible to breast-beating displays of public remorse, and faced with the prospect of infinite prison time, many an innocent person might admit to a crime he or she had not committed, if, by doing so, it hastened that person's release. However, Truscott's pragmatism infuriated the Canadian press. Hitherto staunch defenders of his cause, they now turned on him viciously.

---

*Robert Jackson, *Francis Camps* (London: Hart-Davis, MacGibbon, 1975).

Not all was lost. There was one person whose loyalty never flickered.

Each night Isabel LeBourdais appeared on TV to report that day's events in court and to reaffirm her belief that the Truscott affair was a stain on the nation's conscience. To the end she seemed incapable of grasping the fact that it was her overwrought manifesto—with its distortions, half-truths, outright falsehoods, and absurd conjecture—not Truscott's ordeal, that had dragged the name of Canadian justice through the mud.

This was the view of the tribunal. By an 8 to 1 majority they rejected Steven Truscott's story and opposing medical evidence out of hand, refused his application for a new trial, and reaffirmed the life sentence.

Truscott returned to prison, but not for long. In 1969 he was released, took a new name, rebuilt his life, and married. Almost a decade later he emerged from obscurity to coauthor his own account, *Who Killed Lynne Harper?** Then, after a brief flurry of publicity, he faded once again into the pages of history.

The Case That Won't Go Away

And there he might have remained had not the Canadian legal system been battered all through the 1990s by a barrage of high-profile miscarriages of justice, with compensation payouts soaring to the $10 million level. Suddenly, judicial blunders were hot news. Against this volatile background, in March 2000 a CBC documentary program, *the fifth estate,* decided to revisit Canada's greatest mystery.† Truscott, by now a graying man in his fifties with an unblemished record since leaving prison, appeared and spoke sincerely on his own behalf. Strong support for his claims of innocence came in the shape of Dr. John Butt, ex-chief medical examiner for Nova Scotia, and Dr. Rex Ferris, a veteran pathologist from Vancouver, both of whom questioned the reliability of stomach contents as a means of establishing the time of death.

Said Butt, "I think one thing that we can learn is perhaps what the person ate at the last meal. I don't think it tells you anything precise about the time of death."[22] Ferris agreed. "There really was no valid method used to determine time of death," he said, adding, "I don't think there is any medical or forensic evidence which clearly links Steven Truscott with Lynne Harper's death."[23]

Obviously the 1966 tribunal disagreed. Having heard testimony from

---

*Bill Trent and Steven Truscott (Vancouver: Optimum, 1979).
†"His Word against History" (CBC, March 29, 2000).

internationally recognized medico-legal experts, they overwhelmingly sided with Helpern, Simpson, and even Camps eventually—Penistan got it right. Barring any lengthy trauma that might delay digestion—and Helpern was convinced the assault had been "sudden, short, sharp, and fatal"[24]—all the medical evidence suggested that Lynne Harper died within two hours of eating her last meal.

Perhaps the most startling development in this TV program was the way it pitched another suspect's name into the ring. Safely dead and therefore not a libel threat, RCAF sergeant Alexander Kalichuk was branded an alcoholic pedophile—his criminal record consisted of a $10 fine for two counts of indecent exposure in 1950—who at the time of the killing worked on another air base an hour away, sometimes visited Clinton, and who sold a car shortly after Lynne Harper's death.

While a lengthy report on Kalichuk, retrieved from the National Archives in Ottawa, clearly indicates a troubled personality with many problems, nothing in its nine hundred pages hints at any involvement in Lynne Harper's death. And all that fuss over the car fizzled out once it was learned that the vehicle Kalichuk sold was a canary yellow Pontiac, not a gray Chevrolet.

In the wake of these disappointments came the inevitable accusations of a government cover-up. According to the conspiracists, the military was prepared to go to any length to shield one of its own, although the notion that the RCAF would frame a fourteen-year-old schoolboy in order to let a homicidal pedophile run free is frankly too absurd for comment. Whereas no one knows for certain if Kalichuk was even in Clinton at the time of Lynne Harper's death, Truscott's presence on County Road that hot June evening is a matter of record.

It is not difficult to imagine how such a tragedy might have unfolded. All the evidence suggests that Steven Truscott was desperate to lure some young girl to Lawson's Bush that evening. Had he simply wanted to view some newborn calves, he could have gone alone, or with other boys; but no, he wanted a girl to go with him, making it difficult to avoid the conclusion that sex was uppermost in his mind. For two days he had pestered Jocelyne Goddette to accompany him. When she cried off at the last moment, he happened to meet Lynne Harper.

She was known to be fond of Steven, and would probably have been flattered by the older boy's invitation to see the calves. She might even have seen through this subterfuge. Whether Lawson's Bush was a local hangout for sexually curious youngsters is something only those who lived there could know, but the remarks of one person interviewed by *the fifth estate*

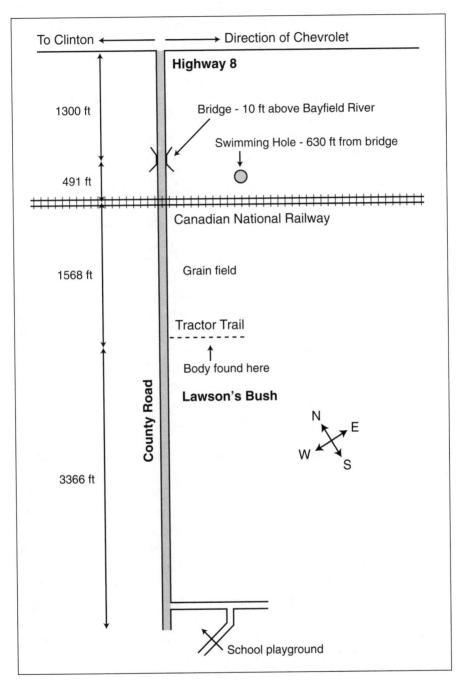

Map of County Road.

program hint strongly at this possibility. Allan Durnin, who was age eleven at the time, said that on the night of Lynne's disappearance, "Butchy George stopped and he said to me, 'Truscott is in the bush with Harper.' And what it meant was that he was in the bush with Harper, trying to do whatever fourteen-year-old boys do."[25] Having inveigled Lynne into the undergrowth, Steven probably attempted intercourse. As noted earlier, decomposition made it impossible to ascertain if this was Lynne's first sexual experience. She might have fought and been killed for her pains. She might not. The almost eerie neatness of the crime scene, with its carefully laid-out clothes, and a lack of bruising on the body, certainly suggest some degree of compliance. The violence may have come later, when the couple were engaged in a fumbling embrace. Suddenly, from a hundred yards away, a young girl's voice cuts through the dense undergrowth.

"Steve! Steve!" It is Jocelyne Goddette, searching for her "date."

Lynne, suddenly furious because Steven has deceived her, begins to struggle, perhaps scream out. To stifle those cries, Steven pushes her blouse up over her face, then yanks it tightly around her neck. He does not mean to kill, but to silence. Instead the girl falls limp. In just seconds the fourteen-year-old schoolboy has become a killer.

Panic-stricken, he hurls a few branches across the body, then dashes to where his bicycle lies hidden, only to gash his leg on the barbed-wire fence. Minutes later the sizzling tarmac of County Road flies beneath his wheels as he pedals back to the base. Behind him a little twelve-year-old body is already under siege from the predators.

Nonsense, said LeBourdais; young Steven didn't have the time in which to rape and commit murder. Even accepting LeBourdais's five-minute time frame—and it is a hypothesis based on the sketchiest of evidence—it was time enough. Strangulation can occur in seconds. In his autobiography *Forty Years of Murder,** Simpson wrote, "Quick tightening of a hand or ligature round the neck can kill like a karate chop—suddenly, of reflex vagal nerve stoppage of the heart action."[26] Many an unintentional strangler has been stunned to find themselves gripping lifeless flesh, unable to believe the lethal vulnerability of the human neck. Rape, too, is often startlingly brief.

Then what about Truscott's doughty courtroom demeanor? LeBourdais protested. Surely here was evidence of a clear conscience? Sadly, she was not the first to be blinded by such apparent guilelessness.

---

*London: George G. Harrap & Co., 1978.

In June 1921, a Welsh youth, Harold Jones, age fifteen, was acquitted of murdering eight-year-old Freda Burnell, who had died from shock after being partially strangled and sexually assaulted at a seed merchant's where Jones worked. It had been Jones's sunny courtroom composure that tipped the scales in his favor. Newspapers and locals rejoiced, united in a refusal to believe that such an atrocity could have been committed by one so young and amiable, and to celebrate, they paraded the beaming Jones aboard a flower-strewn truck through the cheering streets of his hometown. Two weeks later the body of Florence Little, age eleven, was found in Jones's attic. She had been stabbed and raped. This time Jones confessed. At his subsequent trial he also admitted killing Freda Burnell, before he was hurriedly incarcerated for life.

As Jones demonstrated, some children not only have the ability to kill, they also can hoodwink adults. Had it not been for his age, Steven Truscott would now be some half-forgotten sex-killer, and Isabel LeBourdais would not have written the polemic that escalated what had been a senseless teen tragedy into a full-blown "miscarriage of justice." Fortunately, cooler heads were at hand to undermine the rhetoric and expose the pro-Truscott campaign for what it really was—a miscarriage of truth.

Chapter 10

Lee Harvey Oswald (1963)

The Calculating Patsy

No murder in history has been more analyzed than the killing of John F. Kennedy. And no crime has attracted such intense forensic controversy. There isn't a single scrap of information or evidence from that grim November day in 1963 that has not been pored over, turned inside out and shaken, or otherwise scrutinized under every conceivable kind of inquisitive microscope. Depending on who's peering through the viewfinder, upward of sixty different people/organizations have been identified as having had some involvement in the assassination. They range—in no particular order—from the CIA, to Nazi extremists, J. Edgar Hoover, deranged Texas oilmen, the Mafia, KGB agents, gangsters from France, Freemasons, former Israeli president Ben-Gurion, and assorted tramps and transients, all the way up to Lyndon Baines Johnson (who himself always suspected Fidel Castro of having ordered the hit) and dozens of others in between. Little wonder then that by 1998, a CBS poll[1] found that almost 75 percent of Americans did not believe that Lee Harvey Oswald had acted alone, with 68 percent convinced that a government cover-up had taken place. Considering the mind-numbing barrage of conspiracy theories/books/articles/films that have emerged over the decades, one can only marvel that the skepticism rate isn't higher.

Distortions, economies with the truth, and even downright lies have plagued this tragedy from its earliest days, with new misconceptions being added almost daily. Most feed on the public's understandable and deeply-felt yearning for a rational explanation to what was a wholly irrational act. It just didn't seem possible that an attractive and vigorous president of the most powerful nation on earth could be effaced by some Johnny Nobody with a twenty-dollar rifle. Surely, only a plot of labyrinthine complexity could topple this glorious Camelot?

Here, in necessarily brief form, are the facts of the case:

At 12.30 P.M. on November 22, 1963, John F. Kennedy, the thirty-fifth president of the United States, sustained multiple gunshot wounds as he rode through the streets of Dallas, Texas, in a motorcade. Also grievously injured in the ambush was Texas governor John Connally, who was seated in the front passenger seat of the president's open-topped limousine. Immediately, the motorcade sped to nearby Parkland Memorial Hospital, where a team of doctors tried in vain to save the president's life.

Within forty-five minutes of the assassination, a local police officer, J. D. Tippit, was shot to death in a Dallas suburb. Half an hour later, Lee Harvey Oswald, a twenty-four-year-old Louisiana-born warehouse worker and ex-marine, was arrested in a movie theater close to the site of the Tippit murder. At first he was accused of killing only Tippit, but as the evening wore on he became the prime suspect in the murder of the president as well. Throughout the weekend the Dallas police made repeated public pronouncements to this effect. On the fleeting occasions when Oswald spoke to the press, he steadfastly maintained his innocence, and said he would prove it when he was brought to trial.

He never got the chance. On November 24, while still in police custody, Oswald was shot to death by a small-time Dallas hood named Jack Ruby.

Ten months later, a commission charged with investigating the circumstances of that dreadful day and presided over by U.S. Supreme Court Chief Justice Earl Warren, submitted its report. This concluded that Oswald alone had assassinated President Kennedy, and maintained that it had seen no evidence indicating that Oswald and Ruby, together or individually, had been part of a conspiracy to murder the president.

If the intention of the Warren Report was to allay public concerns, it failed miserably. Almost immediately doubts were raised about the legitimacy of its findings, with accusations, murmured at first but increasingly noisy, of a governmental cover-up. As the conspiracy juggernaut gathered steam and the claims and counterclaims became ever more outlandish, even the most minuscule inconsistency was magnified a thousandfold as proof of outside intervention, with most of the accusing fingers being pointed toward Washington.

Such faith in the power of governmental authority is baffling and unsupported by history. If all the executive might and muscle of the Oval Office could fail so dismally in its attempts to cover up a minor burglary at a Washington apartment block, then what price any state being able to conceal a full-blown presidential assassination? Governments just aren't that smart and people aren't that loyal. Somebody, somewhere, would have talked.

While there is insufficient space here to discuss the myriad conspiracy theories currently circulating, we can examine the known (if not undisputed) facts, to see if there was anything at the crime scene to indicate that Lee Harvey Oswald acted in tandem with others when he set out to kill President Kennedy.

Means, motive, opportunity: the three major vertebrae that form the backbone of every significant criminal investigation. Find all three and you've got a strong case; find all three in profusion and you're looking at an almost certain conviction. So let's deal with each in turn and see how they related to Oswald.

Means

President Kennedy was killed by two shots from a high-powered rifle just after his limousine passed the Texas School Book Depository (TSBD), a large brick building that stands at the junction of Houston and Elm Streets, overlooking Dealey Plaza. In the ensuing melee police officers on the spot

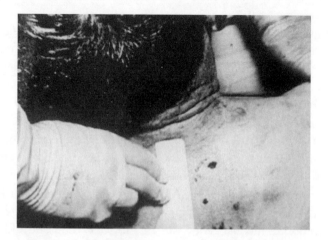

The autopsy of President John F. Kennedy ranks as the most controversial ever performed. (Courtesy of National Archives)

instinctively ran toward the TSBD as being the likeliest source of the shoot-ing. On the sixth floor they found a 6.5-mm, Mannlicher-Carcano bolt-action repeating rifle, complete with telescopic sight. Nearby, a pile of boxes had been stacked by the window to make a kind of sniper's support, giving a clear line of sight down into Dealey Plaza. Three expended rifle shells found on the floor powerfully confirmed the impression of most wit-nesses that the shots had emanated from an upper floor of the Book De-pository. A subsequent serial number check on the Italian-made rifle re-vealed that it had been purchased by mail order on March 12, 1963, by a TSBD employee, Lee Harvey Oswald, using the alias "A. Hidel." From the foregoing, any reasonable jury would have concluded that Oswald had the means to carry out the assassination.

Opportunity

On the day of the killing Oswald was seen carrying a long package, which he claimed contained curtain rods, into the TSBD. Despite an exhaustive search, no curtain rods were ever found at the TSBD—or anywhere else for that matter—and it is logical to assume that the package contained the disassembled Mannlicher-Carcano. Despite some claims to the contrary, Oswald had no alibi for the time of the shooting. He was seen on the sixth floor by a fellow employee about thirty-five minutes before the shooting, and in the second-floor lunchroom by his supervisor and a police officer approximately ninety seconds after the shooting. But for the duration of the assassination itself his whereabouts were a mystery. As any defense at-torney would have been quick to point out, this also meant that no one inside the TSBD could definitely place Oswald on the sixth floor at the critical moment. Even so, he was patently in the building at the time of the shooting, and therefore had the opportunity to carry out the crime.

Motive

Although not required in law, a strong motive is often regarded as the single most telling piece of evidence against the accused. All prosecutors love to be able to turn to the jury and say: "X killed Y because . . ." Here, no such certainty was possible. For all his strident, well-publicized rants against the evils of capitalism and American colonialism, Oswald was never heard to threaten Kennedy personally; reason enough for many to question whether he played any part in the assassination. Anyone tempted to shuffle off down this misguided avenue of thought need only heed the remarks of Tom King, former British secretary of state for Northern Ireland and himself no stranger to the threat of terrorist violence: "If they really want to kill you, they don't ring you up first."[2]

To offset this absence of a publicly expressed motive, prosecutors would have highlighted Oswald's narcissistic and dysfunctional personality. Besides the most highly publicized event in his background—the bizarre 1959 defection to the Soviet Union and his subsequent return to American soil in 1962—there was plenty for the psychiatrists to sink their teeth into. As a teenager Oswald had twice threatened acquaintances with a knife; while in the marines he had shot himself in the arm to avoid an unwanted transfer; and when married to a woman from the Soviet Union, he had repeatedly abused her. A quantum shift in Oswald's level of violence came in April 1963, when, a month after buying the rifle, he attempted to shoot a radical right-wing Dallas activist, Major General Edwin Walker, a fact that didn't become known until after the Kennedy murder.

If Oswald's preassassination behavior indicated a willingness to employ extreme violence, then his conduct post–Dealey Plaza bore all the hallmarks of a mind totally unhinged by the sudden realization of his catastrophic actions. In an instant the loudmouth sloganeer had crossed his own personal Rubicon, only to be pitched headlong into a suicidal haze of self-incrimination.

Within a few minutes of the shooting Oswald left the TSBD, because, as he later put it, foreman Bill Shelley told him that "there would be no more work done that day in the building."[3] (Shelley vehemently denied that this conversation ever occurred.) For such a politically engaged animal—remember, this was a Marxist agitator who could bore the pants off acquaintances with his prolonged and tedious didacticism—Oswald seemed curiously disinterested in the momentous events unfolding right before his eyes. Turning his back on the chaos of downtown Dallas, he returned to his rooming house, grabbed a handgun, then went out again. He appears to have walked aimlessly for about a mile. He had reached the junction of 10th and Patton Streets at approximately 1:15 P.M., when a passing police patrol car, driven by Officer Tippit, drew up alongside and stopped him. Already the Dallas police radio system was humming with the description of a man wanted in connection with the Dealey Plaza shooting, and Tippit thought that Oswald fit the bill. He got out of his car. After a brief exchange of words Oswald abruptly produced a gun and opened fire. Tippit fell, mortally wounded. Two people witnessed the shooting; seven more saw the gunman fleeing the scene. All nine witnesses positively identified Oswald as the killer.

A quarter of an hour later Oswald was seen ducking furtively into a store doorway as the howl of sirens warned of approaching police patrol cars. When the danger passed, he emerged from the doorway and took refuge in the nearby Texas Theater movie house. There, his erratic behavior prompted a call to the police, who converged on the theater in force. After a struggle during which Oswald attempted to shoot yet another officer, he was subdued and arrested.

Oswald did himself no favors in custody. His barracks-room lawyer posturings fluctuated between the pompous and the absurd, and he was caught in a string of ridiculous and unnecessary lies. But he did pull off one masterstroke. That night, while being paraded for the benefit of the press, he famously shouted "I'm a patsy" before he was hustled away. As a serial self-promoter with a proven track record for media manipulation, Oswald knew how to push all the right buttons, and with that single phrase, he laid the cornerstone for a hundred and one conspiracy theories.

Apart from this remark, there is nothing in that day's events to suggest that Oswald was acting in concert with anyone else, or that this was a well-planned operation. Quite the contrary. The whole ghastly episode reeked of spontaneity and makes a nonsense of all those criticisms that Oswald just wasn't "smart enough" to pull off the assassination single-handedly. This was murder ill thought out, ill enacted, and ill covered up.

The Forensic Evidence

Had this case ever proceeded to trial, the main bulwark of the prosecution would have been the incontrovertible fact that Oswald's palmprint was lifted from the murder weapon. The hard-to-reach position of this print—on the underside of the Mannlicher-Carcano's barrel—meant that it could only have been made when the rifle was disassembled, leading to the presumption that Oswald had left the print as he readied the rifle for use. Owing to the roughness of the wood on the stock, it was not possible to check the rest of the rifle for prints.

As Oswald's fingerprints were also found on the discarded wrapper that had been used to carry the rifle, and on two cartons stacked by the window, his presence at the alleged sniper's lair was established beyond a reasonable doubt.

The rifle yielded another important clue. In a crevice between the butt plate of the rifle and the wooden stock was a tuft of several cotton fibers of dark blue, gray-black, and orange-yellow. These fibers were examined by Paul Stombaugh, an agent assigned to the Hair and Fiber Unit of the

FBI Laboratory, who compared them with fibers from the shirt Oswald was wearing when arrested. They matched. "There is no doubt in my mind that these fibers *could* have come from this shirt," said Stombaugh, before adding the caveat "There is no way, however, to eliminate the possibility of the fibers having come from another identical shirt."[4]

Investigators were desperate to know whether Oswald had handled a firearm recently, and for that they turned to the paraffin test. In this controversial procedure, first employed in the United States in 1933, melted paraffin is brushed over the "shooting" hand of a suspect until a thin "glove" is obtained. After cooling, the cast is removed and treated with an acid solution of diphenylamine, a reagent used to detect nitrates and nitrites that originate from gunpowder and may be deposited on the skin after firing a weapon. A positive test is indicated by the presence of blue flecks in the paraffin.

Oswald's results were contradictory. His hands tested positive for the presence of nitrates; his cheeks did not, suggesting that he had handled a gun, but not a rifle fired from the shoulder. Critics of the Warren Commission triumphantly seized upon this as proof positive that Oswald had nothing to do with the Kennedy shooting. Unfortunately, it proved nothing of the kind. Owing to the ubiquity of nitrates in our environment—fertilizers, tobacco, urine, cosmetics, even certain foods—false results to the paraffin test, both positive and negative, are disturbingly common, leading one authority to declare, "The paraffin test is, in fact, nonspecific and is of no use scientifically."[5]

The final piece of evidence linking Oswald—and Oswald alone—to the rifle was the mail order coupon that showed a total cost, including shipping, of $20.95. Document examiners for the Treasury Department and the FBI testified unequivocally that the bold printing on the face of the coupon was by the hand of Lee Harvey Oswald and that the writing on the envelope was his also.

For someone allegedly enmeshed in a conspiracy, Oswald was shouldering an awful lot of the donkey work himself.

Number of Shots

An article of faith among conspiracy buffs is the belief that there was more than one assassin in Dealey Plaza, and that the presidential limousine was caught in a deadly crossfire. To examine this theory in closer detail, we first need to establish just how many shots were fired. Approximately two hundred witnesses were interviewed by the Warren Commission, and, by a margin of almost 18 to 1, they felt that three shots had been fired. Only 5 percent thought they heard more than three. In all likelihood this confusion arose from echoes created by the buildings surrounding the plaza, a phenomenon that probably accounts for the 2 percent of witnesses who thought the shots came from different directions.

In 1979 the House Senate Assassinations Committee (HSAC), established to investigate the murder of JFK, among others, caused a stir when it decided that, based on new acoustical evidence, there was a strong probability that two gunmen were operating in Dealey Plaza. The evidence came in the form of a Dictabelt recording made at the time. Dallas police officer H. B. McLain had been riding a motorcycle in the motorcade, and his microphone was stuck in the "On" position during the assassination and after. According to acoustics experts commissioned by the HSAC, certain impulses on a portion of the channel one Dictabelt recording might have been the effects of rifle fire. Even though these sounds were not distinguishable to the human ear, owing to background noise on the recording, spectrographic analysis confirmed the waveform of the impulses.

It sounded sensational, but the excitement was short-lived. Scientists soon discovered that there had been an "over recording"[6] on the tape, and that the sounds had actually been made later than previously thought. The National Academy of Sciences Committee on Ballistic Acoustics unanimously concluded that "the acoustic impulses attributed to gunshots were recorded about one minute after the President had been shot and the motorcade had been instructed to go to the hospital, and that reliable acoustic data do not support a conclusion that there was a second gunman."[7]

The best evidence—the three expended shells discovered on the sixth floor of the TSBD—strongly favors the three-shots theory.

The Source of the Shots

If there is general agreement on the number of shots, then the source of those shots has always been far more contentious. Each witness's impression of directionality appeared to hinge, in the main, upon where that particular person happened to be standing. Most instinctively felt the shots came from close by: those at the uphill end of Dealey Plaza tended to think the shots came from the direction of the TSBD, while bystanders at the opposite underpass end—by the infamous grassy knoll—thought the shots were fired from there. Since an exhaustive search of the grassy knoll failed to disclose any expended shells, and since the consensus of those present favored a single source of fire, the body of evidence in favor of the TSBD as being the solitary sniper position is overwhelming.

Not only was this view corroborated by witnesses who claimed to have seen a gunman firing from the sixth floor of the TSBD, but also by forensic analysis of the presidential limousine. On the inside surface of the laminated windshield, investigators found lead residue that corresponded with a series of cracks on the outer surface. The physical characteristics of the windshield damage indicated that it was struck on the inside surface from behind, by a bullet fragment traveling at "fairly high velocity."[8] Beneath the left jump seat, which had been occupied by Mrs. Connally, FBI agents found three small lead particles. When compared spectrographically, all the metallic pieces were "found to be similar in metallic composition, but it was not possible to determine whether two or more of the fragments came from the same bullet."[9]

Had a second gunman been shooting from a different direction, logic dictates that some bullet damage to the limousine would have advertised this presence. There was nothing. So either this gunman missed with every shot, or he or she didn't exist.

The Autopsy

Few pieces of evidence are more eloquent than a murdered body. Properly interpreted, they can speak volumes. In the case of death by gunshot wounds, they can tell us the type of ammunition used; the general relationship of the gun to the victim; and, if bullets or fragments of bullets remain in the body, they often can identify the murder weapon. Even allowing for complaints that the removal of JFK's body from Dallas was a flagrantly illegal action—which it undoubtedly was—the autopsy carried out at the Bethesda Naval Hospital in Maryland on the night of the assassination was meticulous and thoughtful, performed by a team who must have realized that their actions would be subjected to the merciless scrutiny of history. Nothing they discovered undermined the lone-gunman theory.

Kennedy had been struck by two bullets. The first hit his upper back and exited through the lower front portion of his neck, clipping the knot of his tie. This projectile then apparently entered the right side of Connally's back, and traveled downward through the right side of his chest, exiting below his right nipple. With its destructive power diminished by every obstruction, the bullet then passed through Connally's right wrist and entered his left thigh, where it caused a superficial wound.

It was the second direct hit that that proved so calamitous. This struck Kennedy a skidding blow to the right rear portion of his head, leaving an entrance wound of 15 mm in length and 6 mm across. The dimension of

6 mm was somewhat smaller than the diameter of a 6.5-mm bullet because the elastic recoil of the skin generally shrinks the size of an opening after a missile has passed through. After causing catastrophic damage to the brain, the bullet exited the head, leaving a gaping wound.

Gunshot expert Colonel Pierre A. Finck, chief of the Wound Ballistics Pathology Branch of the Armed Forces Institute of Pathology, provided the Warren Commission with details of what exactly happens when bullet and skull collide. As a bullet enters the cranial vault at one point and exits at another, it causes a beveling or cratering effect where the diameter of the hole is smaller on the impact side than on the exit side. Based on his observations of that beveling effect on the president's skull, Colonel Finck said, "President Kennedy was, in my opinion, shot from the rear. The bullet entered in the back of the head and went out on the right side of his skull . . . he was shot from above and behind."[10]

Finck was joined in this conclusion by the three doctors who performed the autopsy on Kennedy at the Bethesda Naval Hospital. All the bullet fragments in JFK's skull were right of the centerline, precluding a shot from the right front. They jointly agreed that "the bullet penetrated the rear of the President's head and exited through a large wound on the right side of his head."[11]

Ignoring all those harebrained fantasies that have Oswald's coconspirators somehow gaining access to the president's body and making "alterations" skillful enough to dupe three medical examiners, the trajectory of the bullet wounds suffered by Kennedy and Connally, all of which were in descending pattern, left no doubt that the bullets were fired from above and behind the presidential limousine.

The Ballistics Evidence

Although not of the highest quality, the Mannlicher-Carcano rifle, with its muzzle velocity of approximately 2,160 feet per second, was easily capable of the task in hand, and test firings by FBI specialists proved that this rifle, to the exclusion of all others, had fired the bullets that killed Kennedy and wounded Connally.

It was further linked, though not conclusively, to a hitherto unsolved attempted murder in Dallas on April 10, 1963, when someone had fired a single shot through a window at the home of Major General Walker, narrowly missing his head. Correspondence found in Oswald's belongings left little doubt that his had been the finger on the trigger. Unfortunately, the bullet recovered from that incident was too badly mutilated to permit

Mugshot of Lee Harvey Oswald, taken one day after the assassination. (Courtesy of National Archives)

accurate comparison with the bullets that had killed Kennedy. All Robert A. Frazier, an FBI ballistics identification expert, would say when asked if the Mannlicher-Carcano *could* have fired this bullet was that "the general rifling characteristics of the rifle . . . are of the same type as those found on the bullet. . . ."[12] There were no microscopic characteristics or other evidence to indicate that the 6.5-mm bullet was not fired from the Mannlicher-Carcano rifle owned by Oswald.

The "Magic Bullet"

It's the most notorious bullet in history, and the most cynically exploited. Photographed after being discovered on Governor Connally's stretcher at Parkland Memorial Hospital and called the "Magic Bullet" because of its ostensibly miraculous condition, the Western Cartridge Co. 6.5-mm bullet did, at first blush, seem remarkably unscathed, considering the amount of damage it had inflicted on two human bodies. But, like so much in the Kennedy assassination, the reality is rather more prosaic than the fancy. As another, much less publicized photograph reveals, far from being in pristine condition, the bullet was actually bowed and flattened on its base, with some of its core material gaping out.

This relative lack of damage can be accounted for by the fact that the full metal-jacket bullet went through the soft tissue of Kennedy's neck, tumbled, hit Connally's rib sideways, then his wrist, until finally striking his thigh at a speed too low to further damage the bullet.

Although the single-bullet theory is anathema to those who cling to the belief that Connally was hit by an entirely different shot fired by a second assassin, it does have the benefit of fitting the facts. Let's backtrack for a moment. No one seriously disputes that the first bullet to hit Kennedy was fired from behind and above. As it exited the president's throat on a downward path at an estimated velocity of almost 1,800 feet per second, the bullet was still carrying enormous destructive power. Such a missile doesn't just vaporize—it has to go somewhere. Yet there was nothing in the limousine—no dents, no scratches, no holes—to indicate that it had collided with the vehicle. So unless the bullet suddenly ricocheted through ninety degrees, up and over the windshield, after hitting the president, the only place it could have gone was into the back of Governor Connally, a point emphasized by the Warren Commission. "The relative positions of President Kennedy and Governor Connally at the time when the President was struck in the neck confirm that the same bullet probably passed through both men."[13]

The final and most devastating shot left forty metallic slivers in Kennedy's skull that, when compared spectrographically, matched the bullet fragments recovered from the front of the car. As all the fragments were traced back to bullets fired by the Mannlicher-Carcano, there is no ballistics evidence to suggest that any other firearm was employed in the assassination of President Kennedy.

So if the accumulated evidence that Oswald bought and handled the rifle that killed Kennedy is watertight, what about the oft-repeated argument that this ex-soldier was simply too bad a marksman to carry out the assassination, that he could scarcely hit the side of a barn from twenty paces?

Much has been made of Oswald's alleged lack of shooting prowess. In the marines, despite having twice passed the Corps requirements, he was viewed as a mediocre shot, bad enough for one service colleague to say, "If I had to pick one man in the whole United States to shoot me, I'd pick Oswald. I saw that man shoot. There's no way he could have ever learned to shoot well enough to do what they accused him of doing in Dallas."[14]

But just how difficult was the assassination attempt? The limousine was moving at approximately 11 miles per hour past the TSBD at the time when it is believed that the first shot was fired. Even with a 4× telescopic sight and a slow-moving target, the would-be assassin appears to have missed completely. For whatever reason—an intervening tree branch is the likeliest cause—this bullet was deflected and never found. Then, as the car turned slightly to go down the inclined portion of the street, moving slowly up and to the right across the sniper's field of view, the second and third shots were fired. The final and longest shot, from sniper's window to target, was just 88 yards. Despite the relative ease of the shots, the assassin hit the presumed target (JFK's head) only once out of three attempts, about what one would expect from a nervous, jumpy gunman with mediocre shooting skills. As assassinations go, it was amateurish and inept and makes a mockery of claims that this was the work of a "skilled hired hit man." At that distance a professional killer would have shattered Kennedy's head with the first bullet.

If one accepts this line of thinking, then most conspiracy theories just fall in a hole. What gang of plotters, cunning enough to set up Oswald over several months and not leave a single trace of their own involvement, would pin their faith on such a poor marksman, one who had already botched a previous assassination attempt?

The Zapruder Film

Very few murders are recorded for posterity on film, and had it not been for Abraham Zapruder's presence with a home movie camera at the underpass end of Dealey Plaza, then who knows where the JFK conspiracy buffs might have taken us? If nothing else, the ghastly images captured by Zapruder that day have at least kept the discussion within the realms of the (barely) credible.

When FBI agents examined Zapruder's Bell & Howell Model 414PD 8 mm roll film camera, they established that it ran at 18.3 frames per second. By numbering each frame of the film, it was possible to calculate a time lapse chart of the presidential limousine's brief journey through Dealey Plaza. This proved to be crucial.

Early reports had suggested that the first shot struck the president in the neck, the second wounded the governor, and the third shattered the president's head. But as the film shows, at about frame 160/161 Kennedy looks suddenly to his right, as though distracted by a sudden noise. Within half a second, everyone in the presidential limousine, as well as two Secret Service agents in the following security car, have all turned their gaze sharply to the right. In all likelihood they are reacting to the first shot, which appears to have missed the car completely. This leads us to another area of confusion.

The gut reaction of most witnesses present in Dealey Plaza was that the shots were fired over a period of 5 to 6 seconds, short enough for some to legitimately question whether Oswald was competent enough to fire three rounds in such quick succession (even a skilled gun handler was unable to reload and fire the sluggish Mannlicher-Carcano in fewer than 2.3 seconds). However, the ability to judge the passage of time, especially in stressful situations, is notoriously tricky, and as a frame-by-frame analysis of the film demonstrates, the shooting "window of opportunity" was actually much longer than most people thought.

As we have seen, the film indicates that the first, unsuccessful, shot was fired at about frame 160, well before anyone in the limousine was struck by a bullet. Unfortunately, it is impossible to pinpoint exactly when Kennedy was first wounded. All that can be said with any certainty is that at some point between frame 210, when a road sign obscures the president, and frame 225, when he comes back into view, he has suffered some kind of injury. Frame 225 shows the president hunched over, hands clutching at his wounded throat. Ten frames—or just over half a second later— Connally also exhibits signs of extreme distress after being hit in the back, in all likelihood a delayed reaction to the same shot that wounded Kennedy. Central nervous systems are wholly unpredictable when subjected to sudden trauma, as the Warren Report noted: "In some situations, according to experts, the victim may not even know where he has been hit, or when."[15]

No such ambiguity surrounds the final, killing shot. In frame 313 Kennedy's head explodes. Using simple arithmetic—$313 - 225 \div 18.3 = 4.8$—we now have a minimum time span of 5 seconds between the wounding shot and the lethal shot; time enough, even for the clumsy gun handler, Oswald, to reload, aim, and fire. Remember, he was tardy only by the superslick standards of the marines. Even more importantly, the same arithmetical formula also gives us an approximation of how long the whole ambush lasted: from frame $313 - 160 \div 18.3 = 8.4$ seconds, far longer than most eyewitnesses estimated.

The Zapruder film was doubly informative, as it also allowed investigators to determine whether President Kennedy fell within an assassin's field of vision from the sixth floor of the Book Depository at the times the film says he was shot. On May 24, 1964, a car simulating the presidential limousine was driven down Elm Street, with stand-ins occupying the general positions of the president and the governor. An agent situated in the sixth-floor window tracked the car through the telescopic sight on the Mannlicher-Carcano as the assassin allegedly did on November 22. Films depicting the "assassin's view," made through the rifle scope, paralleled the Zapruder film with uncanny accuracy. At about frame 160 the assassin would have had a clear view of the target traveling slowly below. He fires and misses. As he rushes to reload, his view of the presidential limousine is blocked by the foliage of an oak tree from frames 166 through 210, at which point he fires again, hitting the president in the back. Another five or six seconds pass and he squeezes the trigger for the last time, with hideous results.

This is the most contentious sequence in the Zapruder film. Because the president's head is seen to jerk violently backward from the impact of the final head wound, many have argued that the shot must have been fired from the front. Unfortunately, this is a Hollywood-induced myth. Weaned on a diet of celluloid shoot-outs in which the victim is hurled backward by a bullet's sledgehammer force, it's easy to forget that in real life it just doesn't happen that way. Gunshot victims tend to fall where they're hit. The impact of a bullet may be colossal but it is brief, certainly not enough to cause the fired-out-of-a-cannon effect so beloved of action movie makers.

JFK's nervous system had already been damaged by his first wound, whereby tissue displaced by the bullet pushed into his spinal cord, so that the head shot created instant, massive damage to the nervous system, including stimulation of the nerves in the spinal cord, leading to a neuromuscular spasm.

There is not a single frame of the Zapruder film that supports the notion of a hit squad hidden among the granite and green spaces of Dealey Plaza, waiting to gun down the president as he passed. Nor is there anything else to challenge the belief that all the shots were fired by one person, from behind and above, in a direct line from the "sniper's lair" on the sixth floor of the TSBD.

Had this case come to trial, there would have been enough evidence to send Oswald to the electric chair one hundred times over, and yet, incredibly, the notion persists in some quarters that he was entirely innocent of any involvement in the death of President Kennedy, that he really was a "patsy." Subscription to such a view is to believe that he was a victim of a frame-up so diabolically clever that even Machiavelli might have cooed his appreciation. Not only did the conspiracy ringmasters have the prescience to install this psychopathic loner in a job overlooking Dealey Plaza six weeks before Kennedy's visit to Dallas—well before plans for the Presidential trip were finalized—but they also fooled him into ordering the very rifle that killed Kennedy, forged a photograph of him holding this rifle, arranged for his prints to be left on the murder weapon and the sniper's lair, made sure he had no alibi for the time of the shooting, had him stroll disinterestedly out of the TSBD when all of downtown Dallas was in turmoil, then sat back and toasted their unbelievable good fortune as their "patsy" further tightened the electrodes on his head and leg by gunning down a Dallas police officer for no reason whatsoever, before being taken into custody after attempting to shoot yet another police officer.

It is, of course, a preposterous chain of events. And it never happened. In whatever capacity, whether as lone assassin or part of some sinister cabal—and despite all the millions of words written about this case, there is not a whit of credible evidence to suggest a conspiracy—Oswald was intimately involved in the death of President Kennedy.

He had the means and the opportunity to kill Kennedy, and probably the undeclared motive as well. He was determined to become a "somebody." Nothing in his erratic résumé—his defection to the Soviet Union at the height of the Cold War, his highly provocative pro-Cuba demonstrations on the streets of New Orleans—suggested that this was a man willing to live his days in the shadows. Anonymity was not an option. But to escape the shackles of obscurity requires talent, and here Oswald fetched up short. Inadequate, dysfunctional, resentful of a world that had inextricably failed to appreciate his enormous gifts, he was a caricature of all the "nut with a gun" stereotypes, so it should come as no surprise that he opted for violence. The only shock was in the scale. Embittered frustration and the craving for "recognition" have driven many a psycho to spray bullets around their workplace or across classroom desks; not many choose to squint down a telescopic sight and shoot the president of the United States.[16]

Chapter 11

Jeffrey MacDonald (1970)

Fatal Revision

Nobody sits on the fence when it comes to Jeffrey MacDonald. Everyone has an opinion. To countless millions weaned on the book and miniseries *Fatal Vision*, MacDonald always will be the psychopathic army medic who snapped, slaughtered his family, then went on TV to trade quips with a talk show host. To his legions of supporters—and the number is large and growing—MacDonald remains a martyr, a victim of botched forensic work and suppressed evidence, framed by the army, doomed to a lifetime behind bars because of a governmental conspiracy.

A quick caveat. For this theory to fly, one needs to swallow the notion that the army was ready to throw a highly skilled doctor to the wolves to save a delusional junkie who just happened to be the daughter of a retired lieutenant colonel. Not likely, but when dealing with the extremist wing of the pro-MacDonald lobby, an adult-size portion of suspended disbelief is a big plus.

"People just refused to believe what was there," says John Hodges, nowadays a local police chief, but in 1970 one of the first army detectives on the scene. "They couldn't bring into their minds that someone of his caliber did this."[1]

And it's this observation that strikes at the heart of the MacDonald case. Outraged conspiracists can be a snobbish bunch: they prefer their causes high-profile and, if at all possible, professionally qualified. Deep in their hearts, most would struggle to drum up much enthusiasm for some down-at-the-heels drifter such as Michael Ray Graham, who exited Angola prison to almost universal silence in December 2000 after thirteen years on Louisiana's death row for a murder he patently did not commit, but come up with a hunky, blue-eyed, Princeton-educated Green Beret—*and did we mention he's a doctor*—and you can almost see the chests swell with indignation.

Not that anyone should be surprised, for when it comes to rallying support and soliciting donations, MacDonald always has possessed the Midas touch. In 1979, just before flying back from California to stand trial in North Carolina, he was the guest of honor at a $100 a plate dinner thrown by the Long Beach Police Officers' Association to help defray the cost of his legal expenses!

Read the *MacDonald Defense Update*, an occasional newsletter, or browse the numerous pro–Dr. Jeff web sites, and you might be excused for thinking that none of this would have happened if MacDonald hadn't fallen into the clutches of a scheming, hard-up writer hell-bent on turning this tragedy into a best-seller to revive his own flagging fortunes.

Not so.

However maligned or misled MacDonald might have been by his literary collaborator, nothing in *Fatal Vision* had any bearing on the trial's outcome; indeed, the manuscript had not even been written when the jury decided that this twenty-six-year-old army doctor had slaughtered his entire family.

The details of that dank and drizzly February night in 1970 stunned an entire nation. It began with a phone call to the Fort Bragg emergency services. The male voice, frail and wheezy, mumbled something about a stabbing, then gave an address.

The time was 3:42 A.M.

Within minutes the small red-brick duplex at 544 Castle Drive became an anthill of activity as military police swarmed everywhere. Most felt sickened by what they found. In the master bedroom, Colette MacDonald, pregnant and swollen, lay sprawled on the floor, her pink pajamas dyed crimson with blood, legs apart, head and face smashed to a pulp. Draped across her chest was a bloodstained blue pajama jacket pockmarked with holes. By her side, clad only in blue pajama bottoms, a man lay motionless, head resting on her breast, one arm entwined around her neck. Tossed onto the rug, some way from the bodies, was a small paring knife. Above them, scrawled in blood on the bed headboard, was the word "PIG."

Captain Jeffrey MacDonald groaned. "Check my kids, I heard my kids crying."

Down the hallway the horrors multiplied. In different bedrooms the MacDonald girls—Kimberly, age five, and Kristen, two, had been hacked and battered to pieces.

The entire MacDonald family had been massacred except one, and he clearly needed immediate hospital treatment. Before he was stretchered away, MacDonald jabbered an erratic account of being attacked in the living room by a gang of hippies: two white men, a black man, and a white woman with long blond hair who wore a large floppy hat, high boots, and carried a candle. Throughout the ordeal she'd kept chanting ". . . acid is groovy . . . kill the pigs."[2]

Any mention of murderous hippies was bound to raise the investigative temperature to fever pitch. Just six months beforehand, the Manson "Family" had seared themselves into the American psyche when they slashed a murderous swath through the Hollywood glitterati that left seven dead. They, too, had daubed the word "PIG" in blood at one of their crime scenes. Was the nightmare repeating itself?

William Ivory was already suffering nightmares of his own. As the first army Criminal Investigation Division (CID) officer to arrive at 544 Castle Drive, he urgently needed to restore some semblance of order to this madhouse, more subway station than "preserved crime scene." Upward of a dozen MPs had scurried from jeep to house and back again, all across rain-sodden turf; furniture had been moved; telephones replaced; doors and other surfaces wiped clean; unthinking feet had trampled indiscriminately from room to room. Later, it emerged that someone had even pocketed MacDonald's wallet! How much evidence was either destroyed, contaminated, moved, or compromised, Ivory shuddered to think.

And yet, through all the bedlam and chaos, certain anomalies emerged.

First, the house was still in darkness when the MPs arrived. This struck Ivory as odd. Wouldn't someone dialing for help on the telephone at least want a light on?

And then there was the living room. Considering that MacDonald had fought for his life against three armed male attackers in this confined space, it looked dissonantly pristine. An overturned coffee table, some disheveled magazines, an empty flowerpot that had strewn its contents across the carpet, and a fleck of blood on a pair of spectacles dropped in the corner were the only signs that this small room had staged a life-and-death struggle.

Ivory shook his head and mused.

Nor were all of the peculiarities confined to the crime scene. MacDonald had been rushed to Womack Army Hospital. Apprised of the circumstances surrounding MacDonald's admission and expecting to deal with some high-level trauma, the ER duty physician was shocked to find just a minor chest wound that required a gauze bandage; a bump on the forehead that hadn't broken the skin; and some insignificant stab wounds, none of which needed stitching. The only cause for alarm was a reduction in breathing capability, suggesting the possibility of a partially collapsed right lung, a diagnosis subsequently confirmed.

While being treated, MacDonald elaborated at length on details of the attack. He had been asleep on the living roof sofa, he said, when a sudden anguished cry from Colette jolted him awake. As his eyes adjusted to the pitch darkness, he was able to make out three men standing over him. One, the black man, hit him with a piece of wood; then all hell broke loose. MacDonald fought till he could fight no more, finally succumbing to a torrent of clubbing blows that left him senseless.

Sometime later, when he came to, and oozing blood from his own wounds, he lurched from bedroom to bedroom. Only then did the full horror of the attack reveal itself. In turn he attempted vainly to resuscitate his stricken family; then, drawing on his last reserves of energy, he reeled to the phone and managed to call the emergency services before collapsing next to his wife's mutilated body.

As he oscillated between wild delirium and a chilling lucidity in his hospital bed, MacDonald remained insistent on one point: "Be sure to tell the MPs and CID that I pulled the knife from my wife's chest and threw it on the floor."[3]

At 6:00 A.M. Jeffrey MacDonald finally slid into heavily medicated oblivion.

While he rested, the forensic examination of 544 Castle Drive continued. Outside the back door, investigators found a $2^{1}/_{2}$-foot chunk of wood with, reportedly, two blue threads attached to it. Twenty feet away, beneath a bush, lay a knife and an ice pick. Both had been wiped clean. Whatever else is in dispute in the MacDonald case, by common consent these finds, together with the paring knife found in the master bedroom, make up the quartet of murder weapons. According to the FBI, all originated at the MacDonald residence.

When Colette's body was lifted from the bedroom floor, some blue threads could be seen where she had been lying. This was strange. According to MacDonald, he had rested his blue pajama jacket on top of Colette after finding her. So how had these threads ended up under the body? In all, eighty-one blue threads were recovered from the master bedroom, including two from beneath the daubed headboard. Also, found among the bloodstained bedding was the finger section of a latex glove, the kind that surgeons wear. It had been torn, as though removed in a hurry, perhaps by the person who daubed "PIG" on the headboard.

In Kristen's bedroom a bloody footprint on the floor led away from the body and appeared to have been made by a bare adult foot.

Link to Cult Murders?

But it was what investigators found in the living room that really set pulses racing. Among the disheveled magazines was the latest edition of *Esquire*, which featured a lengthy article about the Manson murders. Had this, detectives wondered, provided the inspiration for MacDonald's tale of a homicidal hippie cult?

Over at Womack, MacDonald was fattening up his story. As the trio of male intruders lunged at him with knives and an ice pick, somehow—he was vague about the details—he managed to slip his pajama jacket over his head, wrap it around his fist, and use it to parry the murderous blows. He blocked some of the thrusts, not all.

Those blows that did strike home left only a pattern of "small puncture marks . . . in vicinity of left chest . . . like pin pricks,"[4] in stark contrast to the wholesale butchery that had been visited upon the other family members. All of which begged the question: why had the frenzied killers allowed MacDonald to live? He was the only witness to their carnage. He could identify them. Yet he had survived with barely a scratch.

Against this undeniably suspicious backdrop the investigation proceeded in a very definite direction. Even before MacDonald left the hospital, CID officers discreetly informed the FBI that they should wind down their search for the elusive hippies—everything now pointed to Jeffrey MacDonald as being the guilty party. At some point they intended filing charges.

Quite why the CID felt such confidence is unfathomable. Put simply, their handling of the case had been disastrous. Nor did their problems end at the crime scene. Vital evidence continued to be either lost or destroyed: MacDonald's blue pajama bottoms, potentially crucial to the prosecution, were burned by an unthinking orderly at Womack Army Hospital; a lab technician, in trying to remove the floorboard that held the bloody footprint, succeeded only in destroying the board completely; unbelievably, skin found under Colette's fingernail disappeared, as did a blue thread from under Kristen's fingernail.

The forensic mismanagement reached all sectors of the investigation. A fingerprint expert, disappointed to find that more than fifty photographs of prints he had located around the house were blurred beyond recognition, went back to the sealed house a second time to reshoot the pictures, only to find that moisture had rendered 80 percent of the fingerprints useless. With that many unidentified prints flying off into the ether, any savvy defense lawyer would be able to drive an eighteen-wheeler through prosecution claims that there had been no intruders.

Offsetting this saga of ineptitude, the CID investigators had uncovered enough inconsistencies to cast grave doubts on MacDonald's version of events. Besides the discrepancies already noted, they included these facts:

1. No fingerprints were found on the paring knife, which he claimed to have pulled from his wife's chest.

2. Neither of the two telephones MacDonald said he used to call for help showed any sign of blood or even fingerprints.

3. Where he said he fell in the hallway after being stabbed was also free from blood.

4. Blue pajama fibers were found all over the house, except in the living room, where MacDonald alleged the fight had taken place.

5. There was a smear of MacDonald's blood on the copy of *Esquire* magazine. Yet, apart from another speck found on the spectacles, this was the only trace of his blood in the room where he claimed to have fought for his life.

It was the blood that spoke loudest of all.

Against all the laws of probability, each family member had a different blood group, which made it possible to track the movement of the

The eerily tidy living room where Jeffrey MacDonald claimed to have fought for his life. (Federal exhibit: United States v. MacDonald)

victims around the house, especially Jeffrey MacDonald. The only two places where his blood showed up in any significant amounts were in the kitchen—in front of a cabinet containing a supply of surgical rubber gloves—and in the bathroom sink, where, investigators speculated, MacDonald had stabbed himself after butchering his family.

Having determined, at least to their own satisfaction, that MacDonald was a mass murderer, investigators were baffled by the seemingly total absence of motive. Here, there was nothing in MacDonald's emotional background to even hint of an impending homicidal explosion, no psychiatric disorder, no reason to indicate why he should suddenly have decided to slaughter his entire family. Confident it could overcome this obvious weakness in their case, in May the army announced that MacDonald was being charged with three counts of murder.

It took the army's Article 32 hearing to reveal the full extent of its own forensic incompetence. Every blunder added weight to defense depictions of the CID as a bunch of Keystone Kops bunglers, a disgrace to serious criminal investigation. Bernard Segal, MacDonald's lawyer, ripped the prosecution expert witnesses to shreds, forcing one crime scene doctor to grudgingly admit that he had rolled Colette's body over, thereby making himself a possible source for those damning blue threads found beneath her. Another humiliation came when Segal established that a hair thought to have come from MacDonald's collar actually originated from the family pony!

As disaster piled on top of catastrophe, the presiding officer, Colonel Warren V. Rock, decided he had heard enough. In his private report, he concluded that the charges against MacDonald were "not true,"[5] and recommended that a greater effort be made to find the gang of hippies. Publicly, the army was less gracious. In October 1970 they curtly announced that all charges against MacDonald had been dropped, without saying more. It had been a humiliating U-turn.

MacDonald immediately filed for an honorable discharge, eager to put as much distance between himself and Fort Bragg as possible. And who could blame him? After all, most people exposed to the merciless glare of a murder inquiry would want to put the tragedy behind them and get on with rebuilding their shattered lives.

Except that MacDonald did no such thing. Ever since the crime he had kept a diary with a view to future publication, noting which magazines and newspapers treated him favorably, how he came across on TV, that kind of thing. Now he drew up a form letter and fired it off to numerous editors, offering himself as a subject for a major article or book.

Jeffrey MacDonald was going after the big bucks.

He might have pocketed a fortune had it not been for a lethal bout of verbal incontinence. In a phone conversation with Freddie Kassab, Colette's stepfather and a hitherto strong supporter of his son-in-law, MacDonald boasted that he and some Green Beret buddies had actually tracked down one of the hippies responsible and *killed him!* It was, of course, utter drivel, and MacDonald would later assert that he had fabricated this claim as a gesture of compassion, to provide Kassab with some "closure"[6] so the poor tormented fellow could get on with his life. All it did was turn Kassab into an implacable and vengeful foe.

Not that MacDonald seemed to notice. By now he was something of a minor celebrity, and on December 15, 1970, fashionably dressed and sporting hip sideburns, he sauntered onto the Dick Cavett TV show and into the homes of millions. In a performance high on personal trauma and utterly devoid of sympathy for his slaughtered family, MacDonald's smug vindictiveness played horribly. After easing him through an account of the tragedy, Cavett asked, "Where are these investigators now who did the original—?"

MacDonald jumped in fast. "Well, most of them have been transferred. It's the army way of handling things. If someone really fouls up, you either give them a medal or you transfer them."

A few in the audience tittered. Cavett shifted uncomfortably in his seat. MacDonald blundered on, joking about the fact that on the night of the murder he had been watching the rival *Tonight* show with Johnny Carson. He also indulged in gross exaggeration. "The army," he said, "didn't consider how I could inflict twenty-three wounds on myself, some of which were potentially fatal."[7]

Throughout the telecast MacDonald had resembled a rookie matador taunting a dying bull, either unaware or contemptuous of the dangers involved. Rubbing the army's nose so publicly in the dirt was not a smart move; nor did his risible claim to have suffered "potentially fatal" injuries endear him to those who knew better. Freddie Kassab, watching the tasteless spectacle at home, alternately shuddered and seethed. Captain Motormouth from Fort Bragg was about to find out that wounded bulls are often the deadliest opponents of all.

Despite the fact that MacDonald was now a civilian, and legally beyond their jurisdiction, the army, spurred on by Kassab, reopened their investigation into the killings. It took time, but by July 1974 enough evidence had been amassed for MacDonald to face a grand jury.

He made little effort to conceal his contempt for the proceedings, spinning a web of lies, half-truths, and contradictions. Finally, the arrogance and hostility, always simmering just beneath the surface, boiled over. In a fury he turned to the grand jury and his interrogators and snarled, "You can shove all your fucking evidence right up your ass!"[8]

Rather than heed MacDonald's heartfelt suggestion, the grand jury returned an indictment that charged him with triple murder.

Big Bucks and Bad Memories

Bringing the case to trial took years. In that time MacDonald amassed a hefty war chest of defense fund contributions from dozens of influential supporters who believed in his innocence. He also underwent a hypnosis session to help his recall of the fateful night. Whether it was a tribute to the hypnotist's skill or MacDonald's extreme suggestibility is unclear, but all of the following details were recalled from a nine-year-old encounter that lasted a few tempestuous seconds in a pitch-black room before he was knocked unconscious:

The taller of the two white men was muscular, had vacant eyes, and a dimple in his chin that measured approximately three-eighths of an inch. The other white man was described as Italian-looking with greasy hair, weasel-eyed, with a thin neck and a weak chin. He'd worn a gray sweatshirt and looked wasted from drugs. The black assailant was squat, thick-lipped, had crinkly hair and flared nostrils. He also had a large neck, probably ". . . a size sixteen or seventeen collar."[9] The girl's hair was blond and straggly; she had a pointed chin and a bump on her nose.

Impressive stuff. Maybe that was why MacDonald looked so relaxed and confident when, in July 1979—nine and half years after the murders—he finally got to face his accusers in a North Carolina courtroom.

The forensic evidence was just as contradictory, just as disorganized and confused as before, except that this time around, the prosecution was packing some extra heat—MacDonald's blue pajama jacket. His claim to have wrapped this jacket around his hands to ward off the attackers' onslaught had been reexamined by an FBI analyst, Paul Stombaugh, and he found that all forty-eight holes in the jacket were smooth and cylindrical and could only have been made if the jacket remained stationary—impossible if MacDonald were dodging a surge of blows.

Even more damaging to MacDonald, Stombaugh then folded the jacket in such a way as to apparently show that all forty-eight holes could have been made by twenty-one thrusts of the ice pick, coincidentally the same number of wounds that Colette had suffered. The implication was stark: forget all that malarkey about the intruders, for whatever reason Colette had been repeatedly stabbed through the pajama jacket by her enraged husband, who then fabricated the hippie story to cover up his actions.

Courtroom attorneys love showy theatrics, and prosecutors Brian Murtagh and James Blackburn decided to stage an impromptu reenactment of the alleged attack on MacDonald. Wrapping a pajama top around his hands, Murtagh tried to fend off a series of ice-pick blows from Blackburn. For his troubles Murtagh received a small wound to the arm, but two telling points had been made. First, all the holes in the pajama top were rough and jagged, not smoothly cylindrical, as in MacDonald's jacket; second, Murtagh was stabbed, albeit not seriously. Yet, when examined at Womack Army Hospital, MacDonald did not have a single defensive wound on his arms.

With just this single item—the pajama jacket—and a highly contentious staging of what might or might not have happened during the fight, the prosecution magically managed to relegate all those other forensic foulups to the dim middle distance. This was legal sleight of hand of the highest caliber.

By contrast, when the defense ignited what they'd hoped would be their big cherry bomb, it turned out to be the soggiest of damp squibs.

For years Helena Stoeckley had been the only tangible link to possible hippie involvement in the MacDonald killings. Age eighteen at the time of the murders and already a veteran police informant, this retired colonel's daughter had alienated her own family and drifted into the Fayetteville drug culture, where acquaintances knew her as a phantasist, a puppy eager to please, someone who had cultivated the knack of telling people exactly what they wanted to hear. Depending on which day of the week it was, Stoeckely was either the wig-wearing, candle-toting hippie goddess chanting "acid is groovy," or she had never even clamped eyes on Jeffrey MacDonald. The solitary constant in her tales was an admission that on the night in question she had been smashed on mescaline, a powerful hallucinogenic. There was nothing unusual about this. Most of Stoeckley's waking hours were spent in a chemical haze.

When confronted with the reality of actually testifying—with its everpresent threat of possible perjury charges—Stoeckley plumped for circumspection and declared on the stand that MacDonald was a complete stranger, someone she'd never met before. This was a tremendous setback for the MacDonald camp, and worse was to come. In the years leading up to the trial, Stoeckley's ravings had been a perennial source of optimism for the defense, but when the judge refused to admit testimony from witnesses to whom Stoeckley had allegedly confessed, MacDonald's chances of acquittal nose-dived.

Then came the biggest setback of all, this time self-administered. Back in 1970 MacDonald had taped an interview with the military investigators, and now he was made to suffer while that tape was played in court. By turns he sounded evasive and indifferent; always he remained arrogant. In one narcissistic passage, MacDonald eulogized himself thus: "I'm bright, aggressive, I work hard. . . . Christ, I was a doctor!"[10] adding, "I had a beautiful wife who loved me . . . what would I have gained by doing this?"[11] At this point a detective had flipped a photograph onto the desk in front of MacDonald.

All the arrogance hissed out of MacDonald when he recognized the young woman's face. The sham was over. Far from being the Norman Rockwell happily married husband, MacDonald was exposed as a serial adulterer. This latest conquest was just the most recent of a string of casual sexual encounters that had littered his marriage. Unhinged by this revelation, he murmured, "You guys are more thorough than I thought."[12]

One juror later remarked, "Until I heard that [tape], there was no doubt in my mind about his innocence . . . but hearing him turned the whole thing around."[13]

The cold, hard print of trial coverage rarely conveys the full story. Physical presence is essential. A jury picks up on nuances; they see expressions, reactions, interplay between principals that never get into the trial transcript, and most of the jurors agreed that MacDonald made a horrible witness. Under cross-examination, he wheedled and shifted and generally did himself no favors at all, shrugging indifferently whenever asked to explain an awkward fact or statement. All of this was avoidable—if MacDonald had availed himself of the right to silence. As prosecutor Blackburn put it: "You pay your money and you take your chance, and MacDonald chose to testify."[14]

Prior to this, one of the jurors, Fred Thornhill, bemused like everyone else by the forensic shambles, had been prepared to give MacDonald the benefit of the doubt. "We all really assumed that he was innocent," he said. But hearing MacDonald testify removed all doubt. "There was a real sense of phonyism [sic]."[15]

In an attempt to boost the weakest part of its case, the prosecution offered MacDonald's sexual peccadilloes as a possible motive for the killings. With Colette pregnant and unappealing, MacDonald, the superstud Romeo, wanted to clear the decks for action. It wasn't much, but it did give the jury something to take with them as they began their deliberations.

They needed just six and a half hours to make up their minds.

Guilty.

On August 29, 1979, MacDonald was convicted and sentenced to three consecutive life terms. Finally, it seemed, the last had been heard of Jeffrey MacDonald. Then, out of the ozone, came one of the strangest, most surreal moments in American jurisprudence.

Killer Sues to Protect His Name

In 1984 MacDonald filed suit against Joe McGinnis, author of *Fatal Vision*, the monumental account of the Fort Bragg murders, claiming breach of contract and fraud. It all stemmed from a contract drawn up between the two men in 1979. In return for a sizable percentage of any profits that might accrue, MacDonald agreed to provide McGinnis with exclusive access, enough for him to produce the definitive account of the case. When it came to money, MacDonald was smart. He knew the commercial worth of his story, and he wanted to pocket his share of the proceeds. Contentwise, the agreed structural spin was that he would be portrayed as "embattled former employee victimized by the monolithic military apparatus."

Except it didn't turn out that way. When *Fatal Vision* was released in 1983, far from receiving the anticipated rose-tinted treatment, MacDonald was stunned to find himself depicted as a psychopathic speed freak who had slaughtered his family in a drug-crazed frenzy. For this perceived betrayal—and McGinnis would later concede that the drug angle was never more than a theory—MacDonald cried foul, claiming that the writer had abused his privileged position.

No one can seriously doubt that McGinnis took MacDonald to the literary cleaners. Published documents show that right up to the guilty verdict, he had continued to offer encouragement and express his belief in MacDonald's innocence, thereby preserving that all-important access while simultaneously producing a manuscript that would forever damn MacDonald in the public eye. It worked. McGinnis got his blockbuster. And a ton of money. He also got slapped with a lawsuit.

The subsequent civil action was long and acrimonious. Expert witnesses argued ethics and the obligations of a journalist to his subject—even if that subject is a thrice-convicted killer—and what part intellectual honesty should play in such a transaction. If McGinnis is to be believed—that he signed the contract believing wholly in MacDonald's innocence, only to be persuaded by the evidence and, vitally, by close personal contact that he had got it wrong—should he then have buried those suspicions and carried on as originally intended? Or would such reticence be tantamount to acting as an accessory after the fact? These were thorny issues that had nothing to do with fact and everything to do with opinion.

When the suit went to the six-member jury, the result was deadlock. Five thought McGinnis had shafted MacDonald; one did not. The impasse ended on August 21, 1987, when the judge declared a mistrial. Rather than go through the whole miserable exercise again, both parties agreed to settle for $325,000, the sum originally requested.

Grasping the opportunity offered by the trial's ambivalent outcome, MacDonald's supporters—every bit as slippery as McGinnis—set about translating what was, after all, a negotiated settlement into a stellar victory, a triumph of shining truth over tendentious journalism. So what if all six jury members later stated that their unanimous opinion of MacDonald's guilt had not been altered one jot by this civil action. The Dr. Jeff bandwagon was on a roll. A new scapegoat had been found—dastardly Joe McGinnis. Now all they needed were a few more discrepancies from that forensic mess at 544 Castle Drive.

Rule number one for diehard conspiracists: Find enough anomalies, no matter how trivial or minute, and pretty soon doubts creep in.

Rule number two: Throw in a juicy chunk of possible prosecutorial misconduct and just watch those doubts take wing and fly.

Right from the outset, it had been a central plank of the prosecution case that there was not a scintilla of physical evidence to support MacDonald's claim that four intruders broke into his house that fateful night. However, as documents retrieved under the Freedom of Information Act have revealed,* not only was the prosecution aware of certain evidence that might support this premise, they also did their damnedest to suppress it.

The most egregious example concerned a synthetic blond hair, some 22 inches in length, found in Colette's hairbrush in the living room where MacDonald said a blond woman had urged on the other gang members. At MacDonald's trial, Michael Malone, a forensic examiner at the FBI crime lab, testified that this saran fiber came from a doll, not a wig, even though the FBI's own documentation clearly stated that saran *was* used to make wigs. (Incidentally, Malone was among a dozen FBI technicians cited in a 1997 Justice Department report that criticized evidence-handling and management problems at the country's premier forensic laboratory.)

During MacDonald's trial, Dr. John Thornton, emeritus professor of forensics at the University of Berkeley, had been a defense witness on the question of the torn pajama jacket. After the verdict he continued to maintain an interest in this case. "The only new evidence the FBI produced at the trial was the folded pajama top, which Stombaugh did, and I think even the FBI was suspicious of that,"[16] he said, adding that a later FBI investigation found that seven of the tears were actually made in the opposite direction to that which Stombaugh claimed. "The pajama-top reconstruc-

---

*See Fred Bost and Jerry Potter, *Fatal Justice* (New York: W. W. Norton, 1995).

tion was the single most devastating evidence against MacDonald," said Thornton. "But directionality meant it didn't happen that way."[17]

Other disturbing discrepancies have emerged. When a single blue fiber was found beneath Kristen's bloodied fingernail, this was used by the prosecution to link MacDonald to the attack, saying that, in fighting for her life, Kristen had snagged her fingernail on her father's blue pajama bottoms. What the prosecutors signally failed to disclose was that also beneath Kristen's fingernail was a brown hair with its root intact, and that a similar but not identical hair, again with its root intact, also was retrieved from under Kimberly's fingernail. Since neither hair could be matched to MacDonald—or anyone else, for that matter—these hairs have been brandished as proof of third-party involvement in the killings, most likely ripped from the heads of their unidentified attackers by two terrified children.

Or does it?

At the time of the frenzied attack that ended her brief life, Kristen was probably fast asleep in a darkened room. What kind of struggle could this two-year-old have mustered? Enough to rip a hair from some unknown assailant's head?

Kimberly's death also came under the forensic microscope. Investigators speculated that she had been awakened by the sound of her parents fighting and had gone to the master bedroom to investigate. There she saw her father in a murderous rage, clubbing Colette to death. When Kimberly attempted to help her stricken mother, she was hurled to one side, crashed headlong into the doorframe hard enough to deposit traces of brain matter, then fell unconscious. Tracking her blood spots along the corridor, it appeared as though she had been carried back to her bedroom, where she was knifed to death. It is hard to imagine some unknown assailant bothering with such niceties.

Yet Kimberly, too, had an unidentified brown hair beneath her fingernail. Where did these hairs come from? One possible solution is that neither hair was connected with the attack, and was merely the accumulated detritus of normal infant activity.

This is, of course, just hypothetical, and in no way excuses the prosecution's shameful lack of transparency in the area of evidence disclosure.

Taken in isolation, it can be argued that MacDonald supporters are blowing up minor discrepancies out of proportion, but all evidence is cumulative. An unaccountable fiber here, a hair there, some contentious blood spots: in the hands of a skilled attorney, they soon mount up.

In the meantime, MacDonald continues to make regular TV network appearances. The broadcasters don't mind paying for the costly satellite feed from Sheridan Correctional Facility in Oregon, because the MacDonald murders are always good copy, and MacDonald himself has never been short of words on the subject, portraying himself as the victim of an army conspiracy to frame him.

Grayer nowadays and much thinner, he likes to emphasize just how dangerous Fayetteville (near Fort Bragg) was at the time of the murders. He, himself, had been the recipient of threats from numerous acid-heads whom he had antagonized. If true, then he was remarkably sanguine about his family's security on that February night. Not only did he fall asleep on the living room couch, knowing that the unlocked back door provided immediate access to the master bedroom where his pregnant wife lay, but he also managed to remain asleep while the intruders entered an unfamiliar darkened house, crept through the master bedroom, along the corridor past the children's bedroom, across the living room where he dozed, into the kitchen, where they selected a couple of small knives and an ice pick, then back to the master bedroom to commence their ritual slaughter.

Only then did MacDonald wake up.

Such a scenario raises one other question: Why, if they were murderously inclined, did the hippie gang come unarmed?

Before her death in 1983, Helena Stoeckley provided one possible answer. The break-in, she said, had started out as just a heavy visit to scare Dr. MacDonald, to intimidate him for his well-publicized antidrug stance, except that the violence got tragically out of hand.

Once again Stoeckley's blurred imagination had come to MacDonald's rescue, as had her unverified claims to have worn a blond wig at about the time of the killings, but that has not stopped MacDonald's supporters from cynically exploiting this pitiable creature.

All the while the search to find the holy grail that will free Jeffrey MacDonald goes on. It might lie in a 1999 application to subject much of the original evidence to the latest DNA testing techniques, the outcome of which is still unknown, or it might lie in some finely argued point of law (certainly a case can be made for claiming that prosecutorial misconduct might have denied MacDonald a fair trial).

A boost to this line of reasoning came in December 1993 when James Blackburn, one of MacDonald's prosecutors, was himself sentenced to three years' imprisonment for embezzlement, faking a lawsuit, and forging a judge's signature. Surely here, at last, was proof that Jeffrey MacDonald had been victimized by a crooked prosecutor?

Of course it proved no such thing; but three decades of straw-clutching do tend to promote a certain desperation. Bad forensics might have helped convict MacDonald, and good forensics might yet set him free, but in the final analysis it was hubris that undid him.

For if ever anyone talked himself into a prison cell, it was Jeffrey MacDonald. Like many individuals with a bloated appreciation of their own worth, he loved to share this fascination with others. Talk, talk, talk, he couldn't stop: his fantastic ramblings at the time of the crime; the grotesque telephone call to Freddie Kassab; the overweening arrogance and blatant lies that littered his taped interview; bad-mouthing a grand jury; hopelessly outfoxed over *Fatal Vision*; his disastrous performance on the witness stand. Just one calamitous utterance after another.

Nobody needed to frame Jeffrey MacDonald; he did an unbeatable job of convicting himself.

Chapter 12

Lindy Chamberlain (1980)

Australia's Forensic Nightmare

According to the poet Robert Frost, "A jury consists of twelve persons chosen to decide who has the better lawyer."[1] Nowadays Frost's wry observation could easily be broadened to include expert witnesses as well; for when it comes to the arcane business of deciding criminal trials, it's not just the quality of forensic evidence that counts, but the manner of its presentation. Expert witnesses come in all shapes and forms; some are fluid and extremely plausible in their delivery, others are less so. While this may have no bearing on the content of their testimony, how it is received by an impressionable jury is quite another matter.

It's no coincidence that the majority of the more polished expert witnesses testify for the prosecution. Most are in its employ, enjoying all the perks of a regular caseload, access to virtually unlimited resources, the latest technology, and those regular, often headline-grabbing courtroom appearances that do so much for the ego and one's professional standing. No struggling along in some small, independent lab for these forensic superstars; they're headliners in the major leagues.

The downside is an expectation of infallibility. Make a mistake, get found out, and your credibility is shattered, perhaps forever, as happened in the case that forensic science would like to forget.

If bungled crime scenes are the biggest bugbear of criminal detection, then bungled science is a mighty close runner-up. Combine the two and you are staring down the barrel of a legal catastrophe. All too often slipshod evidence processing has led to the guilty walking free, while sometimes dooming the innocent to years behind bars. Nowadays, as procedures improve and law enforcement agencies realize just how merciless the glare of media scrutiny can be, the situation has tightened up.

Usually a crime scene is manifestly obvious—a house is ransacked; someone has been mugged; a homicide victim lies motionless on the floor—and the investigation swings smoothly into action. But as the following case demonstrates: what happens if a crime scene isn't regarded as such in the first place?

Every country has its "Crime of the Century." Not many get to stage that crime in the shadow of an undeniable national treasure. Almost at the geographical heart of Australia lies Ayers Rock, or Uluru, its Aboriginal name. A vast slab of redstone, five miles in circumference and rearing up sixteen hundred feet from the flat, arid outback, Ayers Rock is a magnet for thousands of sightseers every year.

Two such tourists on August 17, 1980, were Lindy and Michael Chamberlain. The young couple from Queensland, still in their thirties, had spent the day exploring the great rock, but now, with darkness fast closing in, they were barbecuing their evening meal. Sitting nearby was oldest son Aidan, age six, while just a few yards away, inside the family tent, their two youngest, Reagen, age four, and his sister, nine-week-old Azaria, lay sleeping.

At about 8:00 P.M. Michael was startled by a sudden cry. Lindy immediately rushed to check on her children, only to see a dingo backing out through the flaps of their green and yellow tent, shaking its head, as though carrying something in its mouth. Like a wraith, the wild dog dissolved into the darkness. Although dingoes were by no means uncommon at Ayers Rock, usually they skirted the perimeter of the campsite, scavenging for food and rarely troubling the tourists.

Lindy crawled inside the tent and saw Reagan still asleep on his bed. Just beyond, baby Azaria's bassinette had been tipped over slightly, the bedding askew.

And it was empty.

In a blind panic Lindy ran from the tent, crying, "The dingo has got my baby!"[2]

Within minutes the campsite was teeming with activity. Flashlights pierced the desert darkness as rangers scoured the surrounding area, though most secretly doubted that the baby would be found alive. Other campers had confirmed the presence of dingoes earlier that evening by the campsite, and pointed to dingo tracks around the Chamberlains' tent. All through that dreadful night the Chamberlains waited. Daylight brought no relief. Azaria had vanished.

By this time the tent had been embargoed, though no one had yet thought to photograph the interior. Only later, as rational contemplation gained the upper hand over frenetic impulse, were attempts made to secure the scene. Even to a casual observer there was evidence of blood in the tent, such as might be caused by a wild animal attacking a child. A stain on one sidewall, no more than a few inches from the ground, appeared to be a spray pattern of blood. Again, no photos were taken.

After a few days, when it became clear that Azaria would not be found alive, the Chamberlain family climbed into their yellow Torana hatchback and drove the five hundred lonely miles back to their home in the Queensland mining town of Mount Isa.

Nightmare turned to tragedy just days later when—courtesy of a TV broadcast—they heard that a small pile of baby clothes had been found two and a half miles from the campsite, close to a known dingo lair. The blood-stained jumpsuit, torn and dirty, together with the undershirt and bootees, were identified as those worn by Azaria on the night of her disappearance. Only her lemon-edged matinee jacket—or short coat—was missing. Judging from the way the officer held the clothes in his bare hands, thrusting them toward the TV camera to show the bloodstains around the neckline of the jumpsuit and undershirt, the local police force had learned zilch about evidence processing in the eight days since Azaria's disappearance. Ominously, the broadcast stated that the clothes appeared to have been neatly folded when found.

Public opinion was already beginning to harden against the Chamberlains. In a country where skepticism is chiseled into the national character, many found the concept of "baby-stealing dingoes" just too bizarre to swallow. This was the stuff of nursery rhymes; big, bad wolf, that kind of thing. Besides, no such an incident had ever been recorded. It didn't take long for the doubts to give way to vicious rumors. The most nonsensical had the deeply religious Chamberlains—they were both Seventh-Day Adventists, and Michael was a pastor—slaughtering their infant because she was subnormal, then manufacturing the dingo story to cover their misdeeds.

Behind the scenes there was more than rumor at work.

The police were deeply suspicious. Something about the baby's clothes just didn't sit right. Why was there so little blood on the jumpsuit? Why were the clothes not ripped to shreds? Where was the matinee jacket? Did it exist at all? How could a dingo possibly carry off a ten-pound baby? Why were there no remains found?

Ayers Rock ("Uluru"), the scene of Australia's worst forensic calamity.

Possible answers to at least some of these questions came from canine experts who said that a dingo in the wild gives no thought to storing food for later. It eats what it finds and it consumes everything, the fur on small mammals, even a bird's feathers. Nor would the baby's weight be any deterrent to an adult dingo. Records listed numerous examples of them carrying off a twenty-five-pound wallaby.

None of these revelations did anything to mitigate the vitriolic level of press hostility toward the Chamberlains, particularly Lindy, who had been targeted as the dominant partner. This was wholly due to her personality. Self-possessed and abrasive, not given to weepy displays of public grief, Lindy Chamberlain confounded all newsroom prejudices of how a bereaved mother was supposed to act, and they let her have it with both barrels. Public opinion wasted no time in latching on to media cynicism, with the result that as the inquest drew near, Lindy began receiving death threats.

The inquest opened on December 15, 1980, and dealt mainly with scientific testimony. Dr. Kenneth Brown, a forensic odontologist who had conducted a series of experiments using dingoes, opined that the holes in Azaria's jumpsuit were not made by a dingo's teeth, and were more consistent with cuts made by a pair of scissors or a knife. This suspicion was compounded by an experiment at the Adelaide zoo, in which a parcel of meat wrapped in a baby's jumpsuit was thrown into the dingo enclosure. As the hungry animals tore at the package they left ill-defined tooth marks, nothing like those found on the clothing near Ayers Rock.

Taken in isolation, these findings might have been significant; but when set against the catalog of glaring crime scene mismanagement—the lack of adequate photographs, the physical handling of evidence with bare hands, the absence of thorough forensic testing—they faded into obscurity.

Because of the intense public interest, coroner Denis Barritt took the unusual step of agreeing to televise his verdict. He was terse about the sloppy scientific evidence, rejecting most of it out of hand. Then he turned to the Chamberlains. "You have all been subjected to months of innuendoes, suspicion, and probably the most malicious gossip ever witnessed in this country."[3] It had been a national disgrace, said Barritt, who ruled: "I doth find that Azaria Chantel Loren Chamberlain . . . met her death when attacked by a wild dingo. . . . I further find that neither the parents of the child, nor either of the remaining children, were in any degree whatsoever responsible for this death."[4]

Lindy and Michael Chamberlain left the court, praying they could now put this tragedy behind them and get on with their lives. But no government relishes the kind of public larruping that coroner Barritt handed out, and Northern Territory was no exception.

Government Fights Back

Their response was lightning fast. Dr. Kenneth Brown, whose own reputation had taken a big hit following the inquest fiasco, flew with Azaria's clothing to Britain, where it would be examined by his former lecturer at the London Hospital Medical College, one of the world's foremost pathologists, Professor James Cameron.

A student of the buccaneering Francis Camps, Cameron stepped easily into his teacher's shoes as Professor of Forensic Medicine at the University of London when Camps retired in 1972, soon establishing his own reputation in what was and remains a highly competitive industry. Unlike his predecessor, Cameron was no mortuary slab isolationist. "Delegation" was his watchword. He gathered about him what he called "the team"[5]—all experts in their field—consisting of other pathologists, a forensic odontologist, technicians, and a photographer, all aided by their regular secretary. Cameron saw himself as merely the public expression of their combined efforts—a perception he was at pains to promote—and it was this team he now set to work, charged with deciphering the riddle from the Australian outback.

What they came up with was forensic dynamite, enough for the authorities to apply to the Northern Territory Supreme Court to quash the findings of the first inquest and reopen the case. That wish was granted, and on December 14, 1981, a second inquest was convened.

It is one of the curiosities of human nature that experts tend to gain gravitas with the number of air miles they rack up. This is particularly true in a court of law. For some reason, when a witness has flown across a continent—or in this case, halfway around the world—to deliver his or her testimony, it invariably carries considerably more weight than if the same testimony had emanated from a local lab. With Australian opinion now polarized over the Dingo Baby Case, news that the illustrious Professor Cameron was winging his way from London to testify on behalf of the Northern Territory government provided the anti-Lindy camp with an enormous boost. And yet, despite the fact that it was Cameron who garnered all the headlines, by far the most damaging testimony—as far as Lindy Chamberlain was concerned—came from a source much closer to home.

Joy Kuhl, a forensic biologist with the Health Commission in Sydney, had frequently assisted the New South Wales police in matters concerning blood and other bodily fluids. When the Chamberlains' car had been shipped to Sydney for exhaustive forensic testing, she was the obvious scientist to call. To hear her testimony, the inquest shifted from the Darwin courthouse to a car compound some five minutes' drive away, where Kuhl circled the yellow Torana, ready to publicly reveal what she had found.

Blood!

According to Kuhl, the front passenger area of the car was full of it: on brackets, bolt holes and hinges around the front seats, on the carpet, on a pair of nail scissors found in a pocket of the console between the front seats. And she also had found traces of blood on the zipper of Michael Chamberlain's camera bag, which detectives had long suspected might have been used to transport Azaria's murdered body.

Chipper and confident, Kuhl gave her testimony with immense brio, in the manner of someone delivering canonical truth. The blood, she said, was fetal in nature, and had definitely come from a child under six months of age.

She singled out areas on the console and around the dash where more blood had been found. Most ominous of all was a roof section of the footwell that seemed to show evidence of arterial spray, implying that baby Azaria had been stuffed into the space below the dashboard, still spurting blood.

While defense lawyers looked on aghast and reporters scribbled gleefully, Kuhl laid on a superb piece of theater. Like some stage conjurer, she took a scraping from a piece of vinyl that had been removed from the seat and placed it on a piece of filter paper. When she added a solution of orthotolidine—used to identify the presence of blood—the paper turned first light blue, then brilliant turquoise.

Voilà. Joy Kuhl had proved the presence of blood.

Kuhl left the car compound in triumph, off to test yet more articles belonging to the Chamberlains. She would report back.

When the inquest resumed at the courthouse, it was time to bring on the big beasts of the jungle. And they didn't come much bigger or more impressive than Dr. James Cameron. Consulted by law enforcement agencies from around the globe—though his star was somewhat on the wane in his homeland following some well-publicized miscarriages of justice in which his evidence had been discredited—Cameron still had a knack for nosing out the big headline-making case. Immediately the court fell under his evidentiary spell. Cameron was a professional witness down to his toes, and it showed. Speaking easily and with utter assurance, he described a string of experiments he had conducted on the clothing and a dingo skull that had been sent to him. After dressing a child of similar size and weight to Azaria in a jumpsuit, Cameron said he found it difficult to believe there would have been enough area of skin exposed to allow the neck to be grasped without causing significant damage to the jumpsuit. The damage to the neck of the jumpsuit had been caused, he thought, "By a cutting instrument such as scissors or a knife. But on closer examination, it's more consistent with scissors."[6] The lack of bloodstains, saliva, and tissue inclined him to human rather than canine intervention in Azaria's death.

Cameron next dealt with the contentious issue of whether the clothing showed evidence of dragging. Despite documented evidence to the contrary, Cameron was assuming, like almost everyone else, that a dingo would have been physically incapable of picking up and carrying a ten-pound object, and would have dragged it instead. For Lindy Chamberlain's story to hold water, Cameron said, there had to be rubbing or scuff marks on the clothes, and he couldn't find any.

What he did locate was blood, and by using ultraviolet fluorescent photography he was able to show the court a string of impressive slides that highlighted the areas of blood staining on the jumpsuit. From these he deduced that the blood flowed all around the neck at one time and not from separate areas, as would be expected in a dog attack.

Cameron had paced his testimony like a well-trained actor, and he gauged that now was the time to deliver the climax of his performance.

On the jumpsuit, in the underarm region, he had found evidence of what appeared to be a bloodstained handprint. It belonged, said Cameron, to a young adult. Gasps greeted this announcement. When both

Chamberlain's counsel and the coroner peered into the maze of dark smudges and professed themselves unable to discern anything resembling a hand, Cameron testily pointed out that the untrained eye often had difficulty detecting what was blindingly obvious to the expert.

Cameron finished with a dramatic flourish. Holding the jumpsuit aloft, he announced that "death had been caused by a cutting instrument, possibly encircling the neck, certainly cutting the vital blood vessels."[7] At best, he was saying, Azaria's throat might have been cut, but there was nothing to rule out the possibility of decapitation.

As the conveyor belt of experts continued unabated, it soon became evident that not all the theatrical talent was imported. Nobody in Australia knew more about textiles than Professor Malcolm Chaikin, at the University of New South Wales, and he was wonderfully accomplished at conveying his knowledge. He had been asked to cast his experienced eye over Azaria's clothing. Comparing it to clothing used in the Adelaide zoo experiment and a garment he had himself cut with scissors, Chaikin demonstrated how the perforations in Azaria's clothing most closely resembled his own experiments. He thought that four areas of damage on the jumpsuit showed signs of having been caused by sharp scissors. When he attempted to replicate the damage using a mounted dingo tooth, which he plunged into clothing wrapped around a freshly killed rabbit, at no time did it rupture the material, even when the fabric-covered tooth penetrated the carcass to a depth of one-third of an inch. The implication was crystal clear— a dingo's tooth would tear but not cut.

Bernard Sims, a forensic odontologist and one of Cameron's "team," told the court that while most of his work related to human beings, he did have some experience of dog bites in flesh and clothing, and he believed that a dingo was typical of the dog family in its dental structure. He basically reiterated Chaikin's testimony: the damage to the jumpsuit bore little if any resemblance to that found in the Adelaide zoo experiment, and there was no evidence—either tooth marks or saliva—to show that Azaria's clothing had been in contact with any member of the canine family.

Like Cameron, Sims seriously doubted whether anything smaller than a Great Dane could have carried a ten-pound baby such a distance without any obvious signs of having dragged it along the ground. Sims thought it would be next to impossible for a dingo to bodily carry off the baby as Lindy Chamberlain had said, because its jaws would not extend that far.

As the inquest adjourned for Christmas, the situation was looking decidedly bleak for Lindy Chamberlain. From skimpy beginnings, the state's forensic case was now like a snowball rolling downhill, gaining weight with every revolution, and threatening to turn into an avalanche that would engulf her.

The New Year brought no respite. When the hearing reconvened, Joy Kuhl was back with a whole new batch of results from her latest experiments. Still more blood, fetal in origin, this time found on the camera bag.

One of the Chamberlains' attorneys, Andrew Kirkham, asked Kuhl to clarify the orthotolidine test she had used to determine the existence of blood. He wanted to know if the same reaction could be gained from substances other than blood. She explained that a blood reaction was "very, very distinctive, particularly in the hands of an experienced operator. And I consider myself an experienced operator."[8] Under probing, Kuhl did concede that it was possible to obtain the same reaction from milk and child's vomit, and admitted that she had not tested for these because there was insufficient matter.

"So in the circumstances, you are prepared to assume that it must have been blood?" asked Kirkham.

"I have not assumed," Kuhl snapped back. "I have reported that a blood reaction was obtained."[9] Truculent and feisty, Kuhl stood her ground like Horatius at the bridge, ready to repel all comers.

In essence the state case came down to this: For whatever reason—postnatal depression was the likeliest candidate on offer—Lindy Chamberlain had murdered her nine-week-old baby, probably in the family car, then hid the body in her husband's camera bag, until such time as she could dump it in the desert, before returning to the campsite and fabricating the story of the dingo.

Since so many distinguished forensics experts had cast doubts on Lindy's version of events, was it possible for that many specialists to be wrong?

Unsurprisingly the coroner, Gerry Galvin, thought not, and he referred the case for formal trial. Right from that first evening at Ayers Rock the police had made no secret of the fact that they believed Lindy to be the prime mover in the disappearance of Azaria—enlisting the assistance of her impressionable husband only after the crime had been committed—and the charge sheet reflected this. Lindy was charged with murder; Michael was charged with being an accessory after the fact.

Charged with Murder

In many respects the trial that followed was haunted by a sense of inevitability. With all the evidence out in the open, everyone knew pretty much what to expect. Once again Cameron and his entourage flew in from England, and Kuhl was no less impressive than she had been at the inquest, although she did falter somewhat when asked why she had not presented the original testing plates to the court.

"They have been destroyed," she admitted, adding that it was "standard procedure in our laboratory."[10]

Afterward, sharing a drink with some reporters, Kuhl complained of the way Lindy Chamberlain had stared at her as she gave evidence. "She is . . . a witch. I could feel her eyes burning holes through my back."[11] Then Kuhl beamed as her newfound press friends toasted her birthday.

The defense fought back hard. Their star witness was Professor Vernon Plueckhahn of Melbourne University. Massively qualified and very short-tempered, he took issue with Cameron's testimony on almost every point, particularly that business about the handprint. Struggling to banish the contempt from his voice, Plueckhahn said, "With due respect to Professor Cameron . . . I cannot in the wildest imagination . . . see the imprint of a hand."[12] The so-called imprint, he thought, had been caused by irregular blood flow, nothing more.

Dr. Dan Cornell was equally caustic. As the scientist who had introduced into Australia the crossover electrophoresis screening technique that Joyce Kuhl had used to detect the alleged fetal blood, Cornell was scathing about Kuhl's competence, stating baldly that "she didn't know what she was doing."[13]

The defense experts had plenty of good points to make, but for all their qualifications they lacked the mellifluous polish of their prosecution counterparts, who were seasoned courtroom veterans, well used to playing on the jury's susceptibilities. In the harsh pragmatic world of adversarial justice, evidence-giving is all about salesmanship, and on this occasion the prosecution had lined up the heavy hitters.

On October 29, 1982, both defendants were found guilty. Lindy Chamberlain had lost her daughter; now she'd lost her liberty. Imprisonment for life. Her husband received an astonishingly lenient eighteen-month suspended sentence, supported by a good behavior bond of $500 (Australian).

So it was that with no body, no weapon, and no motive—just a wealth of highly dubious scientific testimony—the Northern Territory consigned Lindy Chamberlain to penal oblivion, hoping against hope that they'd heard the last of that damn Dingo Baby.

They were wrong.

The defense fightback was already under way. A small band of supporters, headed by an obscure lab analyst, was convinced that Lindy Chamberlain had been railroaded. There were no forensic superstars here, just a group of dedicated scientists determined to get to the bottom of this mystery. Les Smith had no great academic qualifications to his name—he worked for a New South Wales food company and held a diploma in applied science—but he knew enough to feel a profound sense of unease at Lindy's conviction.

His disquiet was shared by two colleagues, Dr. Roland Bernett and Ken Chapman, both highly qualified scientists in their respective fields of microbiology and chemistry.

What troubled the trio most was the oft-repeated Crown witness assertion that canids tore cloth and could not cut it. Smith wasn't so sure. Using his own pet collie, Susie, he began a series of experiments to test the truth of this claim. With Bernett and Chapman acting as observers and supervising the photography and written reports, over several months he fed Susie chunks of meat wrapped in a toweling material and meticulously recorded the results. These clearly showed that sometimes Susie tore the fabric, and sometimes she bit through it quite cleanly, leaving those neat cuts that had so convinced the Crown scientists. Try as he might, Smith could not replicate this damage with a pair of scissors.

Now he needed to see the clothing recovered from Ayers Rock. After months of badgering, in September 1984 Smith was finally allowed access to Azaria's jumpsuit. The similarities between the damage on the jumpsuit and the fabric used in his dog experiments were so apparent that Smith found it unbelievable that the Crown experts had ever made the assumption that scissors had been responsible for the damage to Azaria's clothing.

Next, Smith turned his attention to the animal hairs recovered from Azaria's clothing and still preserved on plates. Hans Brunner, with the Department of Conservation in Victoria, and coauthor of the textbook *The Identification of Mammalian Hair*, had originally offered his services to the Northern Territory and was shocked to be rejected. Now he was only too happy to assist Smith. Microscopic examination of the two jumpsuit plates revealed six canid hairs, two human hairs, and one unidentified fiber. No doubt about it, said Brunner, the canid hairs were from a dingo.

But what about the blood? And especially the "arterial spray" on the roof of the front passenger side footwell?

Smith had long been puzzled by this. It seemed such an unlikely location in which to hold a mutilated baby, with blood pumping everywhere. Kuhl's notes had mentioned the presence of sand in the blood spray on the Chamberlains' car, and this set Smith thinking. Could there be something in the vehicle's manufacturing process to account for this spray pattern? He began by examining forty cars of the same model as owned by the Chamberlains.

No less than five displayed a virtually identical spray pattern on the roof of the footwell.

What the Crown scientists had assumed was blood spray turned out to be residue from a sound deadener pumped into the wheel arch under great pressure, which had found its way through a small drain hole in the panel and onto the roof of the footwell. Chemical analysis of the material sprayed onto the wheel arch and the underdash panel revealed it to be Dufix HN 1081, a sand-filled sound deadener used by General Motors in its Holden brand of cars. All five cars, and the Chamberlains', showed the same angle of spray and had the same drainage hole through which the muffling agent passed.

Now it became a question of fathoming just how Kuhl had found the alleged fetal hemoglobin elsewhere in the Chamberlains' car.

Smith was assisted in the reexamination of this evidence by a whole battery of scientists, among whom was Professor Barry Boettcher, the head of biological sciences at Newcastle University. When Boettcher had given evidence for the defense at Lindy Chamberlain's trial, he had fallen victim to "superior qualification syndrome." His testimony that Kuhl seemed not to understand fully the underlying principles of the fetal tests she had used on the Chamberlain family car had earned him a rough ride from the prosecution, and sneering accusations that he was merely a university academic, a dilettante, insulated from the day-to-day world of practical forensic analysis, where real scientists conducted real tests and obtained real results. Throughout the trial this had been a fundamental tenet of the prosecution, the superior qualifications of its experts when compared to those of the defense.

Boettcher had been deeply wounded by the slight, and after the trial had visited the Behringwerke company in Germany which produced the antiserum used by Kuhl. Here he received confirmation that he had used the same batch as Kuhl, thereby validating his criticisms of her results. He also received a signed statement from Behringwerke, saying they could not guarantee that the antifetal hemoglobin antiserum would react only with fetal hemoglobin, as nonspecific reactions could occur when testing denatured adult hemoglobin.

Someone else gravely concerned by Kuhl's methodology was Julie Fry, a laboratory technician with the Western Australia Department of Agriculture. She failed to understand how Kuhl could have concluded, solely on the basis of faint reactions from the orthotolidine tests, that human blood had been present in the samples. In Fry's opinion these tests, especially when conducted on invisible stains, as Kuhl had done, revealed only peroxidase-like activity, which can be achieved from many substances, such as milk and vegetables, as well as from blood. Without further testing it was impossible to conclude the presence of human blood.

Fry's colleagues agreed. Dr. Bob Hosken, a biochemist, started scouring the textbooks for known problems with orthotolidine testing. The answer leaped off the pages at him—a clear warning that traces of heavy metals, especially copper, can stimulate peroxidase activity similar to that of blood. Using an orthotolidine solution, he tested a number of different substances. His test with a sample of copper oxide provided startling results. The reaction was fast and a brilliant blue, the classic blood reaction.

Immediately, warning bells began to shrill.

At the time of the tragedy, Lindy and Michael Chamberlain lived in Mount Isa, the home of the world's largest single mine for silver, lead, and *copper!* Particles of dust from the mine form a semipermanent tower of gray smoke that billows hundreds of feet into the atmosphere and across the neighboring town. In his report Hosken concluded that Kuhl's results showing widespread peroxidase activity was most likely due to the fact that the Chamberlains' car had been covered in particles of copper dust.

In May 1986 Boettcher visited Mount Isa to perform a random survey in the area using the orthotoluidine tests. He tested samples of dust taken from a house key, the wall of the apartment where he was staying, from the door handle of a car, a van, gravel in the Mount Isa mines parking lot, a new chamois cloth after cleaning a Mount Isa car, and dust from the roadside. Each one gave an immediate positive reaction indistinguishable from the reaction to blood.

Quite apart from anything else, these findings would have been sufficient to flag serious doubts about the safety of Lindy Chamberlain's conviction—after all, the single deadliest piece of evidence against her had been the alleged blood traces in the car—but even before this, another, even more dramatic discovery had already restored Lindy Chamberlain to the front pages.

The missing matinee jacket was found.

It had been discovered on February 2, 1986, by a visitor to Ayers Rock, partially buried in the sand, only a couple of yards from where Azaria's jumpsuit and other clothing had been recovered. All along, Lindy Chamberlain had insisted that Azaria was wearing this little jacket when she disappeared. Few at the time believed her. Now it had been found.

Within a week the Northern Territory government released Lindy and announced that an inquiry would be held. The signing officer appended a curious last paragraph. "Although Mrs. Chamberlain's remission is subject to the usual conditions of good behavior, it is not my intention that she be taken back into custody regardless of the outcome of the inquiry."[14]

Cover-up Revealed

The Royal Commission of Inquiry, headed by Justice Trevor Morling, began investigating the Dingo Baby Case in May 1986. It would sit, intermittently, for two years, travel to many different locations, and listen to many different witnesses. The only constant seemed to be the relentless obstructionism of the Northern Territory government. Even at this late stage, it fought tooth and nail to prevent the Chamberlains' car from being independently tested by forensics experts. In the end Morling, infuriated by the continuous governmental stalling, lost his patience and ordered that the vehicle be examined by the Victorian Forensic Laboratory, as the defense had requested.

There, Tony Raymond, a forensic biologist, employed a test sensitive enough to react with even the tiniest amounts of old blood. As the heme molecule in blood is very stable, if the tests in 1981 had shown a positive reaction to blood, they still would.

His results were startling. Raymond could not find a single trace of blood in the Chamberlains' car. Nor, it transpired, was he the first to reach this conclusion. Years earlier, other scientists had also tested the car for blood, with negative results, only for their findings to disappear into a bureaucratic black hole as neither the Crown nor the police saw fit to pass this information on to the defense.

Kuhl's results had to have originated from some other contaminant in the car. Likewise with the camera bag. In a complete reversal of fortunes, it was now Kuhl who came in for an official drubbing, leaving Professor Boettcher, so vilified by Crown prosecutors at the original trial, to savor the pleasures of total vindication.

Les Smith, too, saw all his years of painstaking research pay off, when the footwell "blood spray" was confirmed by the Victorian Forensic Laboratory as nothing more than sound deadener. When Morling accepted the defense's evidence regarding the testing of copper ore, the prosecution's house of cards collapsed in a heap. Science had sent Lindy Chamberlain to prison; now it had freed her.

Morling spared no one in his report, which was submitted on May 22, 1988. Police, park rangers, so-called animal experts, and devious prosecutorial practices all came in for blistering condemnation, but he reserved his strongest criticism for the slipshod science that had condemned Lindy Chamberlain to jail for three years. "I conclude that none of Mrs. Kuhl's tests established that any such blood was Azaria's," he said, continuing, "With the benefit of hindsight it can be seen that some of the experts . . . were over-confident of the ability to form reliable opinions on matters that lay on the outer margins of their fields of expertise. . . . In my opinion, if the evidence before the Commission had been given at the trial, the trial judge would have been obliged to direct the jury to acquit the Chamberlains on the ground that the evidence could not justify their convictions."[15]

Guilt-free, but not suspicion-free, that was the way the Northern Territory government still saw it, as they battled to stop Lindy Chamberlain from claiming compensation for her ordeal. Finally, in 1992, the Court of Criminal Appeal in Darwin awarded the Chamberlains the sum of $1.3 million (Australian).

Significantly, public opinion polls registered a deep resentment among many Australians over this payment. In their eyes Lindy Chamberlain remained a fiend in human form, someone who'd done away with her own child and then concocted a fantastic story about cradle-robbing dingoes. Not possible, they murmured; dingoes don't steal babies.

It was a myth that was well and truly laid to rest in April 1998, when a thirteen-month-old baby, Kasey Rowles, camping with her family on Fraser Island, off the coast of Queensland, was attacked and dragged away by a dingo. Only prompt action by Kasey's father, Alan Rowles, who rushed the dingo, forcing it to release the child and run off, prevented another calamity.

Lindy Chamberlain wasn't so lucky. Not only did she lose her child, but she also had the misfortune to fall foul of a regime utterly convinced of her guilt and seemingly prepared to go to any lengths to prove it. Within hours of Azaria's disappearance the shutters had come down on all explanations save one—Lindy Chamberlain had murdered her own baby. This is a recurring problem around the world. Once the authorities have someone's name in the frame, a dreadful impetus builds; suspicion feeds on doubt, and prejudice feeds on suspicion. Science is supposed to be our safeguard in such a volatile situation, a bulwark against the twin dangers of emotion and bigotry. Here it merely fanned the flames of hate.

What happened to Lindy Chamberlain was nothing less than a forensic lynching.

Chapter 13

Roberto Calvi (1982)

The Curious Death of God's Banker

Murder or suicide? It can be one of the hardest distinctions in forensic science, and one of the most fertile sources of controversy. Despite what some thriller writers and countless conspiracy Web sites would have us believe, it is damnably difficult to murder someone and then attempt to mask it as suicide; usually there is some telltale sign of outside intervention. Over the course of thousands of autopsies, some medical examiners seem to develop a sixth sense for when something is "not quite right." By the same token, they remain stoically unimpressed by hysterical claims of murder when the physical evidence in front of their eyes, supported by a battery of laboratory tests, says otherwise, no matter how bizarre the circumstances that brought that corpse to the mortuary. One of the strangest and most contentious of such incidents—with fallout that lingers to the present day—began in the heart of London on the morning of June 18, 1982.

At about seven o'clock a clerk hurrying north across Blackfriars Bridge happened to glance down at the River Thames and, to his amazement, realized he was staring at the top of a man's bald head. A second glance confirmed that it was, indeed, a body, suspended by an orange nylon rope tied to some temporary scaffolding beneath the bridge. Call it modern-day complacency or a particularly advanced case of British reserve, but after a moment's hesitation the clerk continued on to work. As he explained later, it occurred to him that film crews often shot in that part of London, and he suspected that he had witnessed the setup for some movie scene. Such speculation didn't prevent him from mentioning the incident to a colleague when he reached work a few minutes later. Immediately the emergency services were called.

Since its formation in 1798—which makes it the world's oldest law-enforcement agency—the Thames River Police have been dragging bodies out of London's main arterial waterway, and by 7:30 A.M. one of their high-powered launches was bobbing in the current under Blackfriars Bridge. Peering upward, all on board realized that this was no celluloid fantasy. The man hanging beneath Number One arch was plump and elderly and quite dead. The rope tied to the second rung of the scaffolding and knotted about his fleshy neck was long enough to allow his shoes and ankles to dangle in the incoming tide. When the police cut him down they found stones weighing just under twelve pounds stuffed into the pockets of his expensive gray suit and the crotch of his trousers. There were two watches on the body, a wristwatch that had stopped at 1:52 A.M., and a pocket watch that had run until 5:49 A.M. They also recovered $10,700 in various currencies and a soggy Italian passport in the name of Gian Roberto Calvini.

British immigration records showed that Calvini had entered the country on June 15, but checks with the Italian embassy revealed that no such passport had ever been issued. However, the forged name fooled no one—it bore too many similarities to that of a known fugitive, someone the Italian police had been chasing for more than a week.

At a packed press conference the next day, a Scotland Yard official sent shivers through global financial markets with his revelation that the dead man was none other than Roberto Calvi, chairman of Banco Ambrosiano in Milan, a shadowy character whose tight-lipped dealings with the Vatican had earned him the nickname "God's banker." Just eleven months earlier Calvi had been convicted of fraud and sentenced to four years imprisonment, only to be freed on bail pending appeal.

Now, having fled justice in his homeland, the sixty-two-year-old financier had hanged himself from a bridge in London.

Or had he?

Such a tantalizing confluence of circumstances inevitably meant that the conspiracy buffs circled like vultures. And, to be fair, there was plenty in the life and death of Roberto Calvi to set the antennae twitching.

More than most modern nations, Italy still operates on a system of favors and closet influence, and nowhere is this more true than in the financial markets. Regulation is lax and corruption is rife, with everyone looking for some kind of edge. During the 1970s this usually entailed being an initiate of a Masonic-type lodge called Propaganda Due, or P-2. Since being founded in the nineteenth century, P-2 has drawn its members from every corner of the ruling classes: the government, the military, the judiciary, the financial world, and, of course, from those two bulwarks of Italian power politics, the Catholic Church and the Mafia. Whether P-2's influence is as all-pervasive as its critics imply is unknowable, what matters is the perception, and among the *cognoscenti*, membership in P-2 is prized as a guarantee of success, the golden key that unlocks a future of limitless wealth and influence. Certainly it changed Calvi's life irrevocably—for good and for bad.

After he was inducted into the secretive society at some time in the early seventies—nobody is exactly sure when—his previously lackluster banking career skyrocketed, propelling him all the way to the top job at Ambrosiano, a strongly Catholic financial powerhouse with close links to the Istituto per le Opere di Religione (IOR), the Vatican's own bank. A string of lucrative joint ventures boosted the profits of both outfits and helped cement Calvi's reputation to the point that when his mentor, Michele Sindona, a longtime financial adviser to the Vatican, suddenly found himself facing a prison term for fraud, the protégé was able to step smoothly into the breach.

And for a while all went well. Calvi's carefully spun financial web circled the globe, and Ambrosiano prospered. Its chairman, renowned for his Midas touch and sphinxlike taciturnity, became Italy's most powerful private banker, backed by the Church, the Mafia, and, of course, P-2.

Then the dream soured. For some reason Calvi turned crooked. He began setting up foreign shell companies and funding them with money borrowed from Ambrosiano's overseas branches. Hundreds of millions of dollars flowed into these shell companies, money that vanished forever in a vast, bottomless black hole. As rumors of the swindle began to circulate, P-2 swung into action, throwing a protective blanket around Calvi, strong enough to shield him from judicial inquiries. This security came at a high price: in return, Ambrosiano was forced to make large and dubious "loans" to P-2 members.

With the ice under Calvi's feet getting thinner at every turn, in 1981 the fatal crack appeared. Officials investigating Sindona's Mafia connections were searching the home of Licio Gelli, the P-2 leader, when they found a safe containing dozens of sealed envelopes. These held meticulous accounts of P-2 bribes to judges, industrialists, and politicians right at the heart of Italian government. Also in the safe was a file labeled "Roberto Calvi." Inside were explicit details of how P-2 had manipulated the Italian legal system to save the crooked banker.

As he was bundled off to jail, Calvi loudly maintained his innocence, claiming he had been a dupe, a tool of the real masterminds of the fraud. But nothing could pry apart his lips and make him reveal names. Italy was awash with rumors. Who really ran Banco Ambrosiano? Was it the Mafia? Or P-2? Maybe even the Vatican? Or had they combined in a kind of unholy trinity to bilk investors of millions?

There were still no answers forthcoming on July 20, 1981, when the notoriously uncommunicative Calvi was convicted of fraud and sentenced to four years in prison and a $10 million fine. After surrendering his passport, he was freed on bail, pending appeal.

As the pressure mounted on Calvi, he began to fear for his life. On May 15, 1982, he bumped into a banking colleague, Nerio Nesi, chairman of Banca Nazionale del Lavoro, at Ciampino Airport. Nesi was shocked by Calvi's depressed mental state. "Everyone's exploiting me," the balding banker groaned. "It's too much to bear."[1]

Banker Flees to London

Less than a month later—on June 10—Calvi jumped bail and, with a false passport, fled from Italy aboard the private plane of Flavio Carboni, a hustler with alleged Mafia connections. Carboni himself was not on the flight, but dispatched Silvano Vittor, a small-time smuggler, to act as Calvi's minder. After stopovers in Trieste and Austria, Calvi touched down in London on June 15 and checked into Chelsea Cloisters, a block of service apartments in fashionable South Kensington. As an expression of his intent to avoid recapture, one of his first courses of action was to shave off his mustache.

Carboni, meanwhile, arrived in London with two beautiful young Austrian sisters, Manuela and Michaela Kleinszig, the first said to be his mistress, the other Vittor's girlfriend. All three registered at the Hilton Hotel, while Vittor remained with Calvi.

At this point opinions about Calvi's state of mind begin to diverge. Carboni claims that he saw Calvi on that first night in London and that the fugitive banker seemed relaxed, griping only about the minuscule apartment. Calvi's family remembered things differently. They said that in various phone calls, Calvi's terror was nakedly apparent. Speaking to his daughter in Zurich, he told her to leave Europe and seek refuge in her brother's Washington house, saying, "Something really important is happening, and today and tomorrow all hell is going to break loose."[2] To his wife he expressed an even more ominous worry: "I don't trust the people I'm with anymore."[3]

Tensions mounted in the tiny apartment. All through June 17, Calvi brooded. That night, according to Carboni's later testimony, he invited the disconsolate banker out for a late dinner. When Calvi declined, Vittor decided to take up the offer of a meal, and he left to meet Carboni and the Kleinszig sisters, leaving Calvi to fret alone.

The time was 11:00 P.M.

Nobody would admit to ever seeing Calvi alive again.

At 1:30 A.M., when Vittor returned, the apartment was empty, the banker gone. After waiting up all night, Vittor, by now mightily concerned for his own well-being, fled the next morning to Austria. Carboni also took flight, though by a more circuitous route, first to Edinburgh, then by private jet to Switzerland, where he also went into hiding.

Meanwhile, back in London, amidst all the media hoopla, the body of Roberto Calvi was ready to give up its secrets—or so the authorities hoped.

The autopsy was performed by Professor Keith Simpson. Even at this advanced stage of his career—he was born in 1907—Simpson retained his status as Britain's premier medical examiner, still as lucid as ever when testifying, still unrivaled on the subject of violent death in all its many and horrific forms, still hugely influential. Ironically, for someone who spent most of his adult life in such an emotionally taxing profession, Simpson had chosen pathology precisely because he was squeamish. As a newly qualified medical graduate back in the 1930s he had developed a visceral abhorrence for the physical ailments of his fellow man. He wanted no part of "looking down throats and examining smelly feet,"[4] preferring the attractions of a world in which "my patients never complain. If their illness is perplexing, I can always put them back in the refrigerator, talk over the problem with my colleagues, and come back to it later."[5]

In his early days Simpson had clashed often with the mighty Spilsbury. No two personalities could have been more dissimilar. Sir Bernard, all pomp and patrician arrogance, had founded "Fortress Spilsbury" on a bedrock of infallibity and fought like a demon to keep it that way. Simpson, humbler and by far the superior pathologist, was never afraid to admit to any temporary gap in his knowledge, gladly seeking out and acting on the advice of others if some tricky problem arose. As Spilsbury's powers waned, the mantle was passed, and Simpson soon achieved the professional eminence, if not the public recognition, afforded his predecessor. In what was always destined to be a controversial autopsy, his was the cool head necessary to take some of the calories out of what was becoming a dangerously overheated situation.

Beyond the antiseptic tranquillity of the mortuary, media speculation was running rampant. They had no doubt that Calvi had been "suicided" —gangland slang for a murder made to look self-inflicted—by Mafia assassins. Why, they asked, had Calvi walked four miles to Blackfriars Bridge to commit suicide, when there were plenty of other bridges en route should he wish such a conspicuous demise? Breathless reporters regaled their readers with vivid, often fanciful, accounts of the physical exertions that Calvi would have had to endure in order to hang himself from the bridge. Only someone with the superhuman agility of a comic book hero, they snorted derisively, could have scaled the parapet, descended the twenty-five-foot iron ladder, swung across to the scaffolding, made his way out over the river, tied a rope around an iron pipe and his throat, then launched himself into eternity. Tarzan, maybe; certainly not some paunchy sixty-two-year-old banker, with stones in his pockets and a brick stuffed down the front of his trousers.

As a means of suicide it did sound extravagantly improbable. On the other hand, it was also one hell of a strange way for the Mafia to murder someone, especially in light of their well-advertised preference for a bullet to the head as a means of problem resolution. Such ostentatious theatricality was entirely out of character.

Undaunted, the conspiracists fired right back. Forget the Mafia, they crowed, Calvi had been executed according to a secret P-2 ritual oath that ordains death for traitors in "the ebb and flow of the tides."[6] The claims of Masonic involvement became increasingly lavish. Apparently Calvi's

Number 1 Arch of Blackfriars Bridge in London, where Roberto Calvi was found hanged.

assassins had opted for hanging because the noose was a symbolic representation of the cable tow, a rope that is hung around the entered apprentice's neck during initiation. His pockets were then stuffed with stones because Masonic tradition dictates that when first initiated an apprentice is like a rough uncut stone, and that he is gradually hewn into a "Perfect Ashlar."[7]

Even the choice of Blackfriars Bridge as a gallows took on a sinister significance. After all, it was common knowledge—wasn't it?—that P-2 performed their ceremonies dressed up in black cassocks like Dominican friars, who are known as Black Friars. Moreover, the bridge stood just within the jurisdiction of the City of London police force, a reputed hotbed of Freemasonry, and within sight of the Temple Gardens, also significant, as the Masons are descended from the Knights Templar of the fourteenth century. The height of absurdity was reached with vague, wholly unsubstantiated reports that a member of the British royal family, who was also a high-ranking Mason, had actually participated in the execution ritual!

For those seeking to hedge their bets at the conspiracy craps table there were rumors linking Ambrosiano to the murky world of Latin American arms trading. Since Calvi's death had come just after the end of the Falklands war, might it not be significant that Blackfriars Bridge was painted pale blue and white, Argentina's national colors?

Insulated from all this wild-eyed speculation, Simpson went about his work. He began by microscopically examining every inch of Calvi's body, looking for telltale bruises that would signify violence, and for needle marks that might reveal the injection of drugs, both suggestive of murder. The three-feet-long nylon rope had been fastened around the the neck by a simple noose, and to the scaffolding by two half hitch knots. Simpson paid particular attention to the neck to determine whether Calvi had really hanged himself, or whether he had been strangled beforehand, then strung up on the scaffolding to simulate the appearance of suicide.

Murder by hanging—which is exceedingly rare—almost always produces an impression of the rope in a complete circle around the neck as the noose tightens, whereas suspension suicide leaves an impression of a rope most deeply in the front of the throat and also leaves the mark of the knot, in this case behind the right ear.

In his notes Simpson wrote: "Deep impression of a noose around the neck and upper thyroid level in front and on the left side, rising to a suspension point behind the right ear, a single ropeline weave pattern, with asphyxia petechiae above this level, and internally to the heart and lungs. . . . No other injury to arouse suspicion. No injection marks on the body. . . . No head injury, or bruising. . . ."[8]

On the basis of his findings, Simpson concluded that the impressions he found on Calvi's throat were "those of deliberate suspension and give no cause for suspicion of foul play."[9]

The police were in complete agreement. At an animated press conference, Commander Hugh Moore, the officer in charge, told disappointed reporters, "There are no indications at this stage that it was not suicide,"[10] and he accused the media of overzealousness in pursuing wild stories of ritual murder, and ignoring the wealth of evidence that suggested suicide. As Moore pointed out, Calvi had probably reached the end of his emotional tether and in the final hours of his life his whole world had crumbled around him. Already on the run, hunted by police, his career in ruins, on the last day of his life he learned that the board of Ambrosiano had stripped him of his powers as chairman, and the control of the disgraced institution had been transferred to the Bank of Italy, dooming Ambrosiano to oblivion.* The bank that had been his life for thirty-five years had been snatched from his bumbling, crooked grasp.

Another crushing blow had been news that Graziella Corrocher, Calvi's secretary for many years, had that same day, June 17, jumped out of an office window in Milan, leaving a suicide note that said: ". . . what a disgrace to run away. May he [Calvi] be cursed a thousand times for the harm he has done to everyone at the bank, and to the image of the group we were once so proud of."[11]

Calvi, Moore said, was disgraced and humiliated, aware that in a matter of days an Italian court would almost certainly consign him to a prison cell for four long years. Unable to face such shame, he had toppled over the brink of self-destruction.

Oblivious to police protestations that all rational conclusions pointed to suicide, the press vied with each other to produce the most fantastic theories. They turned to the notorious "missing briefcase." According to Calvi's wife, this briefcase was "always at his side."[12] Now it was missing. Newspapers posed these questions: Were there secret documents in that briefcase that the sinister P-2 or the Mafia, or even the Vatican, did not want to be revealed? Did someone in those organizations fear Calvi was about to crack and decide to silence him before he did so?

When Carboni and Vittor were eventually tracked down, they did nothing to quench the conspiracy flames. Each claimed that his sudden flight from London had been motivated by self-preservation, that having

---

*The following year Banco Ambrosiano collapsed, with debts of $1.3 billion.

helped Calvi to escape and having him vanish inexplicably, they now feared for their own lives.

In a lather of excitement, reporters converged on the inquest on July 23, 1982, pens poised to record the expected scandalous revelations about Calvi's links with P-2, the Mafia, and the Vatican.

It didn't work out that way.

The inquest could not have been more humdrum. More than three dozen witnesses, mostly police officers, gave evidence more designed to fulfill legal requirements than pad out newspaper columns. Carboni and Vittor weren't even there, testifying through depositions. It was left to Simpson to provide the only concrete evidence as to how Roberto Calvi met his end. To the palpable disappointment of the press gallery, Simpson merely said, "There was no evidence to suggest that the hanging was other than self-suspension."[13] The evidence of the rope mark around the neck and the absence of bruises elsewhere on the body, he said, precluded either an earlier strangling or external violence of any kind.

In the day-long hearing there were no earth-shattering revelations about P-2 or the Mafia, no hints of murder, nothing the press could sink their fangs into. Even the dead man's own brother put a damper on speculation when Dr. Lorenzo Calvi, in a move that earned him the undying enmity of other family members, sent a statement confirming that Roberto had previously exhibited suicidal tendencies. "In July 1981, he cut his wrists and swallowed a quantity of tranquilizers in a moment of desperation."[14]

When testimony ended at 7:00 P.M., the coroner, Dr. David Paul, called a brief recess before making his summation. Yes, it would have been awkward for Calvi to hang himself from that scaffold, but it would have been equally difficult to murder him there. How, Paul asked, could a man of Calvi's weight have been carried down that ladder, then across the gap to the scaffolding, then hanged, without "sustaining some marks upon his body of that carriage across that rather awkward scaffolding?"[15]

Dealing with the possibility that Calvi might have been brought to the bridge by boat, Paul noted that the Thames current was swift at that point, and asked, "Could a boat be handled with sufficient skill so that it could maintain its position beneath the scaffolding while a heavy man such as Mr. Calvi was supported upright and then suspended from a rope?"[16]

In conclusion, Paul explained to the jury the three verdicts they could reach: suicide, murder, or an open verdict. "The open verdict may seem like a super open door to scuttle through if you are in any difficulty about returning another verdict," he said. "Let me tell you that this was not, never has been, and I hope will never be a convenient, comfortable way out."[17]

At ten o'clock that same night, the jury foreman stood up in the small coroner's court and announced, "By a majority verdict the jury has decided that the deceased killed himself."[18]

Fury in Calvi's Homeland

In Italy the verdict was greeted with howls of derision, especially by the bulk of the Calvi family. Their concerns were twofold. First, there were the religious implications: Calvi was a Catholic and, as a declared suicide, might be denied burial in consecrated ground; and second, there was a hefty $3 million life insurance policy that might not be paid in the event of suicide.

Even though in March 1983, three Italian forensics experts, having studied all the case material, conluded that it was "probable"[19] that Calvi had taken his own life, the clamor continued to crescendo. Unease also was building in London, so much so that on March 29, 1983, the Lord Chief Justice, Geoffrey Lane, took the highly unusual step of quashing the suicide verdict, on grounds that the coroner had rushed through the case with unseemly haste, and that he had steered the jury away from an open verdict. A new inquest was ordered.

It got under way on June 13, 1983, under a new coroner, Dr. Arthur Gordon Davies. Once again Simpson was the star witness, except that this time he faced a withering assault. It came in the diminutive person of George Carman, Queen's Counsel (Q.C.), the "Silver Fox." Having been retained by the Calvi family, Carman soon showed why he was fast becoming the most feared cross-examiner in British courtrooms as he skillfully maneuvered Simpson into agreeing that it would indeed have been easier for men in a boat to hang Calvi from the scaffolding than for Calvi to have accomplished the task himself. Carman then suggested a reason for the absence of bruises. Might not Calvi have been immobilized by an untraceable drug such as ethyl chloride, injected in some hard-to-detect area— under the hair on his head, for instance? Simpson was obliged to concede the possibility, but added, "I didn't think of such remote matters when I examined the body. . . . If there was a pinprick on the scalp, for example, I would defy anyone to find it."[20]

Donald Bartlett, of the Thames River Police, who took down the body, also felt the weight of Carman's brusing persuasiveness, being forced into an admission that, when the Thames was high, an experienced boatman could have steered a small craft under the bridge, and from it hanged a body to the scaffolding.

This time Vittor did testify, and said that Calvi, on the last day of his life, had been racked by guilt and depression, and had told him of Banco

Ambrosiano's decision to fire him "through clenched teeth."[21] He resisted loudly Carman's claims that he was "trying to hoodwink the jury."[22]

In the hands of a skilled advocate the tiniest inconsistency can multiply into a mass of doubt, and Carman was an undoubted master of this technique. Building his case brick by brick, he piled up the suspicions: the bizarre circumstances of Calvi's death, the known Mafia connections, Carboni's mysterious trips to different cities all over Europe before and after Calvi's death. Cunningly, he even offered the jury a conspiracy theory of his own. "I have never sought to suggest that if Calvi was murdered, Carboni was involved. Whether he was aware of it is another matter entirely."[23]

It all sounded wonderfully sinister, and it produced the desired result. On June 27, 1983, the jury returned an open verdict. Calvi's relatives heaved a sigh of relief and thanked the court for "removing the stigma of suicide"[24] from the family name. In Britain at least, the mysterious death of Roberto Calvi had been officially declared "unsolved."

This cleared the way for Calvi's body to be removed to Italy, where it was interred in the family's private chapel in the cemetery at Drezzo, a picturesque hamlet near a hilltop that marks the border with Switzerland.

But the furor surrounding his death refused to abate. In August 1984 it piqued the curiosity of Dr. Thomas Noguchi, the former chief medical examiner of Los Angeles County, and so-called "Coroner to the Stars." While in England attending a conference, Noguchi met with the first coroner, Dr. Paul, who revealed some hitherto unpublicized details about Calvi's death.

"His shoes were muddy," said Paul. "If he committed suicide, how did his shoes get muddy when he allegedly stepped off a city street and onto a ladder to hang himself? . . . Even more important, his suit was wet up to the armpits. If you're going to hang yourself you don't jump into deep water up to your shoulders."[25]

Paul theorized that Calvi left his apartment that night with no preconceived ideas of suicide; this explained the four-mile journey to Blackfriars Bridge. But as he rambled in the darkness, despondency took hold and climaxed on the Victoria Embankment, where he surrendered to impulse, and made not one but *two* attempts on his life that troubled night.

The first came when, in a suicidal despair, Calvi stuffed his pockets with stones and jumped into the Thames to drown himself. But the deceptive shallowness along this stretch of river fooled him—a long, rocky spit juts above the water just a hundred yards or so east of Blackfriars Bridge—and he only fell in up to his armpits. Thwarted in this attempt, he staggered aimlessly along the muddy bank until reaching the bridge. There, dangling from the scaffolding, was an orange nylon rope, perhaps left by a boat owner to tie up, and he conceived his idea of suicide by hanging, leaving his damp suit and muddy shoes as the only clues to his previous, unsuccesful drowning attempt.

A glance at the local tide tables confirms that Paul might be on to something. On the night of June 17, high tide occurred at 10:59 P.M., about when Calvi was allegedly last seen alive. Eight and a half hours later, when he was found with his feet dangling underwater, low tide had come and gone at 6:25 A.M., and the river was in flood once again. Assuming that Vittor is telling the truth, this would allow the profoundly depressed Calvi several hours to make his way to the vicinity of Blackfriars Bridge and there undergo a terminal personal crisis before deciding to kill himself.

Compare this to Carman's assassins-in-a-boat theory. For this to work, Calvi had to be drugged and brought to the bridge by boat at high tide or close to it, then suspended from the scaffolding and left to die. Even at this hour, the bridge, river, and Embankment are rarely deserted—this is, after all, the center of London—so any gang of killers adopting such bizarre methods would have been running an extraordinarily high risk of discovery. And for what purpose? To put the fear of the Great Architect in the hearts of fellow Masons? It doesn't sound probable. It also should be remembered that this was one of the shortest nights of the year. The weather was fine, and the first streaks of daylight would have darted along the river as early as three-thirty, further reducing the window of homicidal opportunity.

Still not satisfied by the verdict of the second inquest, the Calvi family pursued their arguments in the courts of Italy. After years of clamor, on December 16, 1998, the body of Roberto Calvi was exhumed, and sent for examination to the Institute of Forensic Medicine in Milan, where a three-man team, including the anthropologist Luigi Capasso, who had been working on the five-thousand-year-old remains of Oetzi, the iceman whose mummified body emerged from an Alpine glacier, would finally attempt to decide, once and for all, whether Calvi was murdered.

After two years of painstaking research, their preliminary report was issued in 2001. Using DNA techniques unavailable at the time of Calvi's death, Professor Antonio Fornari, one of the team, said: "The experts have found nothing that contradicts the theory that Calvi was murdered, and several elements that support it. They include a deep and significant bruise on his right wrist which is not compatible with a mark made by a rope, which shows that someone probably seized him by the wrist before he died."[26]

Although it defies belief that after eighteen years of decomposition, Calvi's body should now show "a deep and significant bruise" that Simpson could have missed, Calvi's son Carlo Calvi, a Montreal businessman, welcomed this latest development. The forensic team's findings had justified his difficult decision to have the body exhumed, he said. "We must do all we can to show that he did not commit suicide. My mother has been trying for the past 18 years to give a meaning to my father's death. There is a wealth of evidence and testimony that my father was murdered on Mafia orders. He also had many enemies within the Vatican."[27]

Meanwhile, in Italy all manner of Mafia hoodlums are trying to cut deals with the authorities, all claiming to have the "inside story" on how Calvi died. One mobster turned informer, Marino Mannoia, alleges that Francesco Di Carlo, a Mafia heroin dealer, strangled Calvi with his hands before hanging the body from the bridge to make the death appear like suicide. He says Di Carlo was acting on orders of the Mafia, to whom Calvi owed tens of billions of lire.

Such a claim should be taken with a very sizable grain of salt, since manual strangulation leaves entirely different marks on the neck to those impressed by suicidal hanging, and it is unthinkable that someone with Simpson's matchless experience of violent death would have been fooled by such a clumsy subterfuge.

This was a point emphasized by Renato Borzone, a lawyer for Carboni, who is still under investigation by the Italian authorities. Borzone has gone on record as stating his belief that, ultimately, an Italian court will rule that Calvi committed suicide. As he put it, "The first British inquest was carried out by leading authorities and was beyond reproach."[28] Even so, nothing is likely to quench the flames of this particular controversy.

We may never know what demons, either psychological or physical, accompanied Calvi as he made his last troubled walk along the Victoria Embankment on that fine June night in 1982, but it is clear that the man himself had some sense of his impending doom. Just before his disappearance, in a rare interview with a journalist, he spelled out his fears. "In this sort of atmosphere, any barbarity is possible. A lot of people have a lot to answer for in this affair. I'm not sure who, but sooner or later it'll come out."[29]

The world is still waiting.*

---

*In April 2002 the finalized Italian forensic report was released. To no one's surprise it concluded that Calvi had *been murdered*. Reasons given included an absence of dust under Calvi's fingernails from the stones in his pockets, and the fact that his shoes displayed no traces of copper-lined zinc from the scaffolding poles. As of this writing, it is still unclear whether any criminal charges will ensue.

Chapter 14
Colin Stagg (1992)

Mind Games

In the hierarchy of murder victims—and, sadly, it does exist—Rachel Nickell was top drawer. She was young and blond, sexy and smart, a devoted mother, pure media gold. Best of all, as far as Britain's press and TV were concerned, her infectious zest for life had been captured on a home video, scenes from which would make for some of the most enduring news images of 1992. Women like Rachel aren't supposed to get murdered. So when this attractive twenty-three-year-old was butchered like a hog in broad daylight on Wimbledon Common, it created a double-edged sword for the investigating detectives. On the one hand, it aroused immense media fervor and public sympathy—always useful in any murder inquiry—but the trade-off came in high expectations. This was one murder that the police had to solve; failure was not an option.

Situated in southwestern London, Wimbledon is a place of stark contrasts. In the minds of most people, British and non-British alike, its name is indelibly associated with the game of tennis. For two weeks each summer the world's top players converge on this leafy suburb to do battle in the world's top tournament. The fans come to cheer their favorites, gorge themselves on strawberries and cream, and, hopefully, top up that tan. Few venture far beyond the ivy-clad confines of the All-England Tennis Club, with fewer still realizing that just a few hundred yards to the west lies a stretch of open heathland that in places can be startlingly isolated. Covering more than eleven hundred acres, Wimbledon Common is a rare oasis of greenery in the concrete sprawl, a place of recreation for joggers, cyclists, even horse-riders, and all of it crisscrossed by a bewildering maze of paths.

Roger McKern knew the Common well. As an actor employed at the local theater, he regularly cycled to work, and at ten-thirty on the morning of July, 15, 1992, he happened to spot a woman standing by a bush, playing with a toddler. With her long, flaxen hair and lithe appearance—she had done some part-time modeling—Rachel Nickell cut a memorable figure. She looked happy, too. Not a care in the world as she frolicked with her son, Alex, and pet dog, Molly.

Just fifteen minutes later, a retired architect, Michael Murray, was also strolling along that same path. Where it meandered through a woodland glade, he spotted what he took to be a young woman sunbathing. But as he drew nearer he realized that this was no rustic idyll. Rachel Nickell was dead. Not just murdered, but the victim of a frenzied attack, stabbed forty-nine times, and nearly decapitated. Her blue jeans had been yanked down and she had obviously been cruelly assaulted. It was an appalling scene, made ghastlier still by the sight of little Alex, clinging to the lifeless body, sobbing over and over again, "Get up, Mummy, get up, Mummy."[1] Soon the infant would be overwhelmed by shock, and he didn't utter another word for twenty-four hours.

Ironically, Rachel had only recently started visiting Wimbledon Common. She'd been drawn by the prospect of finding a refuge from the unwanted attentions of perverts who had pestered her nonstop in the parks closer to her home in Balham. On this, the last morning of her life, she had parked her silver Volvo near a landmark windmill on the eastern side of the common and then set out with Alex and the dog.

Somewhere she met a killer.

Had she been a prostitute or some kind of vagrant, her murder would scarcely have rated a media mention, but as we have noted, Rachel had a rare appeal, which guaranteed that her death was emblazoned across every front page in the land. Adding spice to the journalistic pie was the fact that little Alex had witnessed every second of his mother's death.

But just what had the boy seen? He was barely two years of age and could hardly be expected to furnish any worthwhile description. Gradually, under the gentlest coaxing possible, Alex revealed what he knew. All he could say was that his mother's assailant had been white and male.

It wasn't much to go on for the fifty-four detectives assigned to the case, but it did tally with a description given by two witnesses, who noted a man acting suspiciously in the general area at the time. One had seen him by Curling Pond, and described him as age twenty to thirty, about five feet, ten inches tall, with short brown hair and wearing a white shirt and blue jeans. A few minutes later, another woman saw a man, also wearing a white shirt and jeans, some way off washing his hands in a stream, but she wasn't able to see his face. A third witness came forward with a sighting of yet another man—this time tall with very long hair—at about 10:30 A.M., also in the vicinity of the murder.

Either the killer had been fantastically careful or unbelievably lucky, for the crime scene failed to yield a single worthwhile clue: no blood except that of the victim; no DNA samples; no fiber or hair evidence; and no weapon, which was thought to be a single-edged sheath knife. Considering the frenzied nature of the attack, such a total absence of clues had to be considered downright freakish.

Certainly the detective in charge of the operation, John Bassett, thought so, which is why, without any leads to go on, he decided to enlist the help of the latest weapon in the war on crime.

Few forensic developments have generated so much heat and controversy as offender or psychological profiling. Some swear by it, others swear at it; the only certainty is that it's here to stay, and likely to become ever more ubiquitous. At its core is the premise that criminals leave "psychological clues" at the scene, and that by sifting these clues, skilled interpreters can piece together a picture of the likely culprit. Unfortunately, this means that much of this analysis relies on retrospective data—who committed a similar type of crime and how, etc.—and while solutions to crimes of the present may be suggested by crimes of the past, the savvy profiler has to remain aware that mankind's capacity for evil innovation is seemingly infinite. Common sense, observation, and geographical considerations play as big a role in this process as does psychology, for only by studying all aspects of the crime can the profiler hope to be successful.

Screenwriters have been quick to seize upon profiling's dramatic possibilities, usually portraying the profiler as a man—and it always does seems to be a male—with almost supernatural deductive abilities, a modern-day Sherlock Holmes able to provide hapless detectives with an uncanny insight into the mind of the unknown criminal that they are pursuing. In the real world, things aren't quite so pat. Profiles are merely an indication of what type of person *might* have committed a certain crime. Nothing is written in stone, and nothing can be taken for granted.

When it comes to offender profiling in England, the best-known name is that of Paul Britton.

Profiler Takes Center Stage

A chubby ex-police cadet turned psychologist, Britton has been a lightning rod for controversy throughout his career. Heralded by some, particularly his former police colleagues, as an ardent ally in the fight against crime, Britton also has had to weather a storm of criticism from other mental health professionals, deeply suspicious of his undeniable fondness for the limelight and alleged readiness to take credit that rightfully belongs elsewhere.

On this occasion, Britton visited the crime scene, made his observations, studied all the available literature, crunched a few numbers, and came up with the following:

> The offender would be of not more than average intelligence and education. If he is employed he will work in an unskilled or laboring occupation. He will be single and have a relatively isolated lifestyle, living at home with a parent or alone in a flat or bedsit.
>
> He will have solitary hobbies and interests. These will be of an unusual nature and may include a low-level interest in martial arts or photography.
>
> He will live within easy walking distance of Wimbledon Common and will be thoroughly familiar with it. He is probably not currently a car user.[2]

There is nothing here that would surprise anyone with more than a passing interest in crimes of violence—bright, well-adjusted people with solid family backgrounds rarely kill strangers—but it was what Britton said next that raised so many eyebrows and concerns in the psychological/psychiatric community.

> After examination of the source material, I am of the opinion that the offender has a sexually deviant-based personality disturbance, detailed characteristics of which would be extremely uncommon in the general population and would represent a very small subgroup within those men who suffer from general sexual deviation.[3]

These were murky waters indeed. According to Britton, the murderer belonged to a microscopic percentage of even the pervert population. The subtext was obvious: find a local man who fit the profile, and there was a strong likelihood that you'd have your killer.

One expert who profoundly disagreed was Robert K. Ressler, the god-father of psychological profiling in the United States. By coincidence he had been in London just weeks after the crime, and had been given access to some of the case papers. Unsurprisingly, his initial profile matched Britton's in almost every respect, but he couldn't accept the notion that the killer had to come from a single, infinitesimal sector of the population. "The form of deviancy reflected in this particular investigation is not unusual," said Ressler. "There are many, many introverted, inadequate, and some actually deviant young males in our society that constitute a significant number."[4]

But Britton would not be deflected. He had set out his stall; now it was the job of the police to find someone who slotted into his theory. They tapped into *Crimewatch,* a BBC crime fighting program with a good record of generating tips from the public. When computer-generated pictured of both men seen on the Common were shown, the response was enormous, with more than eight hundred calls logged. One name cropped up no fewer than four times.

Colin Francis Stagg was a twenty-nine-year-old odd-job man, who lived alone on a housing estate less than a mile from Wimbledon Common. He had already been interviewed by the police after a neighbor reported that on the morning of the murder, Stagg had appeared strangely excited and talked nonstop about the killing, displaying an apparent knowledge of the crime that appeared to transcend what was then in the public domain.

When detectives visited Stagg at his apartment, a warning sign on his front door proclaimed: "Christians keep away, a pagan dwells here."[5] Inside they found numerous pornographic magazines, books on the occult, an altar, magic symbols and candles, a veritable cornucopia of the bizarre. They had to almost rub their eyes in disbelief. It was as if they had walked right into Paul Britton's profile.

Stagg was arrested on September 19 and taken to the Wimbledon police station. He made no secret of his fascination with the Common, telling detectives that he visited it almost every day of his life. His knowledge of the terrain was encyclopedic; he seemed to know every clump, hollow, and pond by its traditional name. He admitted exercising his dog on the Common on the morning of the killing, but not for long. A headache had forced his early return, whereupon he had fallen asleep on the sofa, watching TV. He had been awakened, he said, by the sound of a police helicopter circling overhead, and later learned about the murder from a local storekeeper.

As detectives delved further into Stagg's private life, the previous tinkle of alarm bells reached a loud shrill when he frankly admitted that he was still a virgin, owing to the fact that he "just couldn't get it up."[6] But he resolutely denied any involvement in the death of Rachel Nickell; he denied being the man seen near Curling Pond; and he denied ever washing his hands in the stream, which he said was polluted and stank.

The more Stagg talked, the more he laid himself bare, quite literally. He admitted that a few days after Rachel's death, he had decided to indulge his passion for nude sunbathing on the Common, much to the annoyance of a female passerby who claimed that he had deliberately flashed her. On September 22 Stagg pleaded guilty to indecent exposure and was fined £200 ($300). As he ran grinning from the court, Stagg jeeringly made a few indecent gestures at the gaggle of waiting reporters and photographers who now hounded his every movement. Had he known what the investigative team was cooking up for him, he wouldn't have appeared so cocky.

At the request of Detective Inspector Keith Pedder, Britton studied tapes of the police interviews with Stagg. Pedder wanted to know what, if anything, in the tapes could clear Stagg as a suspect. Britton thought for a moment, then made the following extraordinary statement:

"Well, for example, if he said he'd been happily married for two years and has a baby, that wouldn't be consistent with the killer. Or if he had a long-term occupation that required a high level of intellect; or if he had demonstrated that he had had successful, stable relationships with women. These things would eliminate him."[7]

Psychobabble, misplaced arrogance, just plain foolish, or worse—it didn't matter. With these few words Britton had shifted the burden of proof from the prosecution to the defendant, pitching eight hundred years of English common law out the window. So far as the police were concerned, if Stagg didn't eliminate himself from the inquiry, they weren't prepared to do it for him. After two fruitless months, the investigation into the death of Rachel Nickell had reached critical mass. It was now about to lurch into a realm of Orwellian manipulation.

A middle-aged woman named Julie Pines had approached the police with a letter written by Stagg to herself, in response to an ad she had placed in a lonely hearts column in November 1990. Clearly Stagg wasn't a man to mince words. In only his second letter he had described a fantasy in which he lay naked in a park, masturbating, only to be surprised by an attractive woman who, far from being repelled, offers herself sexually to him. So disgusted was Ms. Pines by Stagg's coarse effrontery that when he made a follow-up phone call, she told him to clear off and threatened to contact the police if he pestered her. Curiously, though, her self-proclaimed distaste didn't extend to the point of destroying the letter, which she had kept for two years.

As Britton pondered the letter's contents he fielded inquiries from detectives about the viability of mounting an operation in which a female undercover police officer might befriend Stagg and fuel his sexual fantasies in hopes that he might reveal his complicity in the death of Rachel Nickell. Britton felt that, carefully handled, the idea had merit. The detectives were delighted. Without a jot of hard evidence against Colin Stagg, if he was a killer, it was their only chance of putting him behind bars.

"Operation Edzell," as it was called, got under way on January 19, 1993. This was the day when police officer Lizzie James (not her real name) wrote to Stagg, saying that while visiting her friend Julie Pines, she had stumbled across his erotic letter. Describing herself as "blonde . . . [and] attractive," Lizzie told Stagg his letter had interested her "greatly"[8] and that she yearned to begin a relationship.

The effect on Stagg was electric. By return mail he wrote back, a panting missive that outlined his attributes and interests—once again his fondness for nude sunbathing took center stage—and expressing the hope that they would get to know each other intimately.

After this opening salvo, the exchange of correspondence became increasingly sadomasochistic, dripping with promises of domination and humiliation from both sides. Just a few letters in, egged on by Lizzie at every turn, and Stagg was up to full speed, scraping the crevices of his febrile imagination for ever more terrifying images. "I am going to destroy your self-esteem," he pledged in one particularly lurid letter. "You will never look anybody in the eyes again."[9] When Lizzie cooed her pleasure, Stagg was ecstatic, convinced he'd found the girl of his dreams. For their part, the police felt certain they'd found the man of their nightmares.

Paul Britton hatched the psychodrama; his police admirers wrote the script; and Lizzie mouthed the words. Under Britton's stage management, the sexual titillation was ratcheted ever higher. Lizzie began throwing out dark hints about something bizarre in her background, something that she found profoundly exciting. Stagg's interest sizzled. But Lizzie remained teasingly coy as the next major hurdle—telephone contact—was overcome. This occurred for the first time on April 28, 1993. After the initial awkwardness, which was only to be expected, Stagg began to open up, complaining that his neighbors were spreading rumors about him. When Lizzie feigned ignorance and asked why, Stagg promised to reveal all in his next letter.

Before this arrived, a high-level meeting of Scotland Yard officers was convened to review Operation Edzell and determine whether it should continue. After glowing presentations from Britton and Pedder, there could only be one verdict: carry on.

As promised, in his next letter Stagg explained that he had been arrested for the murder of Rachel Nickell. In the clearest terms possible he denied any involvement with the crime, a stance he maintained unswervingly throughout the twenty-eight-week duration of Operation Edzell. Whenever the subject of Rachel Nickell was broached—and Lizzie made sure it never strayed far from the front burner—Stagg's protestations of innocence remained constant.

As the phone calls became more intimate, Britton decided that the time had come to initiate face-to-face contact. He picked what he thought was the most propitious date possible: Stagg's birthday. On May 20 the couple met in Hyde Park, for the first time. A cordon of undercover police officers encircled the meeting place in case anything went wrong. And Lizzie was wired for sound.

A rainstorm nixed the hoped-for picnic and forced a retreat to a small café, where the couple began chatting easily. After a while Lizzie alluded to the dirty little secret in her background. Now that they had met, she said, she felt confident enough to unburden herself. She blurted out that she had been involved in the ritual murder of a young woman and a baby, an experience she found so exciting, so fulfilling, that she felt unable to be truly intimate with another man unless he, too, had taken a human life.

Stagg was torn. For the first time in his life an attractive woman was showing interest in him, and he sensed that here—on his thirtieth birthday—was the opportunity he craved, a chance at last to lose his virginity. All he had to do was say the word. Instead, he meekly mumbled that he could not meet Lizzie's grotesque expectations, and restated his innocence of the murder of Rachel Nickell. The overwrought exchange lasted the better part of an hour, at which point Stagg rose to leave. As he did so, he handed Lizzie another letter.

When read later, its contents generated sparks of anticipation among the police officers. Not only did it contain one of Stagg's more highly charged fabrications, a nasty piece of fiction in which two men—Stagg and another—work out their frustrations on a compliant and joyful Lizzie, but also there were repeated references to a bloodstained knife that became a tool of sexual titillation.

Detective Inspector Pedder turned to Britton in awe. "It's just like you said . . . you just seem to know so much about Stagg."

"Not at all," Britton replied modestly. "I know things about sexual deviancy," adding, "You're looking at someone with a highly deviant sexuality that's present in a very small number of men in the general population."

"How small?" asked Pedder.

"Well, the chances of there being two such men on Wimbledon Common when Rachel was murdered are incredibly small."[10]

Paul Britton, the most celebrated clinical forensic psychologist in the land, had spoken. Henceforth, the investigation of Rachel Nickell's murder would thunder along with just one aim in view: Convict Colin Stagg.

The Net Tightens

Events moved as Britton predicted. Stagg, still unwilling to admit to Rachel's murder but more desperate than ever to impress Lizzie, decided to enhance his sexual prospects by inventing another "murder." In yet another phone call to Lizzie he claimed to have strangled a girl in the New Forest—a story that was investigated and found to be a total fabrication—but to Stagg's delight, Lizzie seemed thrilled by this revelation. Her letters now fantasized about having violent sex with him, during which he produces a knife. At their next meeting Stagg suggested that she should spend the weekend with him, and together they could visit Wimbledon Common.

This was the connection the police had been waiting for. But when Lizzie began pumping Stagg about the Rachel Nickell murder, he threw a wall up around himself, adamant that he had had nothing to do with her death. Lizzie even tried to turn his complaints of constant police harassment to her advantage, saying excitedly, "I wish you had done it—knowing you had got away with it. . . . Screw 'em!"[11]

The war of nerves dragged on, week after week—Lizzie constantly trying to wheedle a confession out of Stagg, dangling the carrot of sex; Stagg refusing to grasp the bait. Finally, Lizzie laid her cards on the table, announcing that she could never sleep with Stagg unless he was the killer. "If you're not that man, and not done that kind of thing [sic], you'll never *ever*, *ever* be able to fulfill me!"[12]

Such aggressive provocation, when later made public, caused a storm of protest in the mental health community. Dr. Glenn Wilson of the Institute of Psychiatry was particularly scathing. "Offering sex if he could prove that he had that kind of disposition . . . this we would call 'operant conditioning.' The layman might well call it brainwashing."[13]

But all this criticism was in the future. For now, Britton was omnipotent, and he decided to raise the stakes. This time Lizzie's script called for uncompromising bluntness. "Quite frankly, Colin, it wouldn't matter to me if you had murdered her—I'm not bothered. In fact, in certain ways, I wish you had because it would make things easier for me."[14] She used this veiled reference to her own alleged "ritual murder" of a woman and a baby, to force Stagg's hand.

"If only you had done the Wimbledon Common murder," she whispered. "If only you had killed her it would be all right."[15]

"I'm terribly sorry," replied Stagg, "but I haven't." Time and again he emerged as a timid puppy, eager to impress, terrified of rejection. When Lizzie talked of the enjoyment she derived from hurting people, he mumbled pathetically, "Please explain as I live a quiet life. If I have disappointed you, please don't dump me. Nothing like this has happened to me before. Please, please tell me what you want in every detail."[16]

For all his reticence, Stagg did let slip one curious anomaly. During the course of one meeting, according to Lizzie, he prostrated himself on the ground to demonstrate the position in which Rachel Nickell's body had been found. And he described details of the position of her hands, gleaned, he said, from a police photograph produced at his interview. Yet detectives present at that interview were emphatic that the single photograph shown to Stagg could not possibly had revealed all the details he had recounted. Such knowledge could only mean one thing: Stagg must have been there.

Frustration now gripped the investigative team. Most genuinely believed, through gut instinct if nothing else, that Stagg was involved in the Rachel Nickell murder, but he was showing no sign of cracking; the hoped-for confession had failed to materialize; and it was becoming increasingly difficult to justify a campaign of psychological warfare that had already dragged on for seven months and cost the taxpayers £1 million ($1.5 millon). It was time to put up or shut up.

On August 17, 1993, Colin Stagg was rearrested and charged with the murder of Rachel Nickell. Video footage of the subsequent interview shows him to be remarkably composed throughout the ordeal, arms folded, one leg crossed over the other, even when "Lizzie James" suddenly enters the interview room and reveals her true identity. Stagg didn't flinch; whatever sense of betrayal he may have felt, it was well concealed behind a string of dully repetitive "No comments."

And it remained that way throughout the thirteen months that Stagg spent behind bars before his case came to trial. In that time his defense team got to learn the full extent of the war of nerves that had been waged against their client, and it also gave them time to muster their own experts from the world of forensic psychology. What they found appalled them. Dr. Eric

Shepherd, a forensic psychologist with thirty years' experience of helping both prosecution and defense cases, described the tapes as "unremitting pornography. . . . I was personally and professionally outraged to learn that a forensic psychologist had guided its construction."[17]

One of the earliest exponents of offender profiling in Britain, Professor David Canter, echoed the criticism. At the same time he questioned the motives of those behind the scenes. "I'd like to know a lot more about Lizzie James and whoever gave her guidance . . . 'cos it tells us something very curious about their view of the world and the fact that they could invent something like that."[18]

As the arrows came winging in from every direction, Paul Britton found himself increasingly isolated. Scarcely anyone in the psychological community could be found who supported his tactics. And there were doubts, too, about the legality of the operation. Britton has always claimed that Operation Edzell was sanctioned at the very highest level, not just Scotland Yard, but also the Crown Prosecution Service (CPS), where lawyers had run a critical eye over the operation and declared it to be admissible in court.

Then came a bolt from the blue.

In March 1994 a judge in Leeds, Yorkshire, threw out the case against Keith Hall, a businessman standing trial for the murder of his wife, Patricia, who had vanished in the spring of 1992 and who had never been found. In remarkably similar circumstances to those of Operation Edzell, an undercover policewoman (also called Liz) had befriended Hall with the intent of eliciting information, preferably a confession. Hall became smitten and asked the woman to marry him, but she demurred on grounds that his missing wife might turn up at any time, whereupon Hall blurted that there was no chance of Patricia returning, as he had strangled her and incinerated the body. Even though this confession was recorded on tape, the trial judge held that it should be excluded under the provisions of the Police and Criminal Evidence Act (1984), and Hall was acquitted.* Despite this setback, the CPS decided that the circumstances between the Hall and Stagg cases differed sufficiently for the trial to proceed.

It was probably the worst decision ever made in the hundred-year history of the CPS (until 1986 called the Director of Public Prosecutions Office).

---

*Interestingly, the trial judge, having excluded the confession, took the unusual step of authorizing its public release; an action interpreted by many that, while constrained by law into excluding a confession, he believed it really should have been admitted.

Legal Wrangle Rages

For five days the trial judge, Henry Ognall, listened to arguments from both sides. Having familiarized himself with the seven hundred pages of correspondence, telephone conversations, and taped meetings between Stagg and Lizzie, he reacted favorably to defense submissions that Stagg had been the victim of a sophisticated sting operation designed to take advantage of a lonely, vulnerable young man desperate to lose his virginity. In asking the judge to exclude the evidence of Lizzie James, defense counsel William Clegg, Q.C., also argued that Britton's evidence should be similarly excluded, as the two were inextricably linked.

For the prosecution, John Nutting, Q.C., struggled manfully but knew he was holding a weak hand. All he could offer was the contentious point about the position of Rachel's hands, which the defendant had described as being in a palm-to-palm, prayinglike attitude, submitting that such details could only have been known to the assailant and therefore amounted to a confession.

Clegg countered by saying that Stagg had been mistaken. Rachel's hands had actually been crossed at the wrist. And he had mangled other crime scene details, including its actual location, and his repeated assertion that Rachel had been raped, when she had not.

When Justice Ognall decided that he didn't need to hear Paul Britton's evidence, the prosecution feared the worst. And so it proved. On September 14, 1994, the judge dismissed all charges against Stagg and set him free. Lambasting the undercover operation as "misconceived," the judge said it betrayed "not merely an excess of zeal, but a blatant attempt to incriminate a suspect by positive and deceptive conduct of the grossest kind."[19] As for the position of the hands, he said, "I am not satisfied . . . that this one very small piece of evidence does amount to, or is properly capable of, being viewed as . . . a confession."[20]

He remarked that any "legitimate steps taken by the police and the prosecuting authorities to bring the perpetrators to justice are to be applauded—but the emphasis must be on the word 'legitimate.'" Here, there had been "a skillful and sustained enterprise to manipulate the accused, sometimes subtly, sometimes blatantly."[21] Much of the harshest criticism was reserved for the role played by Paul Britton, who Ognall described as "the puppet master . . . who pulled the strings."[22]

It was a devastating attack from the bench. Few could remember a senior British judge being so critical of the police or their methods; but it should be remembered that Ognall had the benefit of other testimony, which went largely unreported in the media. Britton's theory that the killer was a sexual deviant so rare that anyone who fitted his psychological profile was almost bound to be the killer received short shrift from Professor Canter, who told the court that the killer seemed like a brutal man who went berserk when he met resistance—a personality type that is all too common.

In a later interview, Canter amplified his concerns. "Clearly the operation was very poorly conceived, because it led to an innocent man being pilloried, and it led to the police being made a laughing stock and being severely criticized by a judge."[23]

But this was not how the right-wing British press—always noticeably coy on the tricky subject of miscarriages of justice—viewed it at all. A tidal wave of coded articles flooded off the presses, all with just one aim in mind: to convince the reader that Stagg, the weirdo loner, had dodged justice only on a technicality. This was journalism with a nudge and a wink. Detectives, too, dropped broad hints that if only the jury had been allowed to hear all the evidence, then a conviction would have been a formality.

What they pointedly neglected to mention, however, was that it was the total absence of solid forensic evidence or conclusive eyewitness testimony against Colin Stagg that had forced the prosecution to rely on the grubby sweepings from this failed honeytrap. They had nothing else. Like the police before them, the CPS were driven by desperation into the clutches of Paul Britton, and they paid the price.

Many were also troubled by the fact that, in choosing to place Stagg at the center of this grotesque psychodrama, Britton was pushing at the margins of what is acceptable in a civilized society. Had Stagg turned out to be a homicidal maniac, the inflammatory nature of Lizzie's goading, the incessant teasing and tormenting, the almost terminal sexual frustration may well have spun him out of control, with potentially catastrophic consequences.

The fallout from this disaster was spectacular and far-reaching. Colin Stagg continued to be hounded by the media, and he continued to have low-level scrapes with the law. In May 1995 he was given twelve months' probation, after threatening a man with an ax on Wimbledon Common. In mitigation, Stagg stated: "Every time I go out, because of who I am, there is a threat of being beaten up."[24]

In September 1999 former detective inspector Keith Pedder—he had retired in December 1995 after 20 years' service, on grounds of ill health, suffering from depression and exhaustion—found himself facing trial for an unrelated charge of corruption. Again, the case was thrown out by the judge, who ruled key evidence "unfair."[25] Pedder felt that he had been scapegoated by disappointed police force czars, and mused publicly on how much this prosecution had to do with his stated intention to write a book, giving his side of the Rachel Nickell murder investigation. To date this book remains unpublished.

In April 2001 it was revealed that the young police woman, "Lizzie James," had received £120,000 ($180,000) compensation for the stress she endured during this farcical procedure. The woman, already retired in her thirties and in receipt of an "enhanced pension," claimed that she suffered from post-traumatic stress disorder as a result of what was allegedly a highly dangerous operation, though in light of the fact that Colin Stagg was not the murderer, the danger was clearly more perceived than real.

Oddly enough, the least affected in all of this was Paul Britton, who appeared to wash his hands of all responsibility, saying, "I don't manage investigations. How could I? I advise the police. What they do with the information is up to them."[26]

His reputation, especially among the police, remained as high as ever, and he continued to be consulted and to give his opinions until 1997, when a serious road accident damaged his spine and effectively ended his career. No longer fit enough to tramp around crime scenes, he switched to teaching hostage negotiation and personal safety.

But the specter of Colin Stagg continued to haunt him. In September 2000 the British Psychological Society announced it would hear accusations of professional misconduct against Paul Britton over his handling of Operation Edzell. As of this writing, the outcome of this investigation is still pending.

Psychological profiling remains a valid and valuable tool of criminal detection, but it should never be viewed as a substitute for solid, hard evidence. Perhaps the last word on this emotive subject is best left to renowned clinical psychologist Dr. Gisli Gudjonsson, who examined Stagg three times for the defense and found no trace of the deviancy that Britton claimed to be able to discern by remote control: "The police must take a more critical view of profiles that are produced, and not use them as something that is foolproof and a hundred percent accurate, because clearly they aren't."[27]

Chapter 15

O. J. Simpson (1994)

When Money Met Science

In the vast majority of criminal trials, the single defining difference between the prosecution and the defense is one of resources. Nowhere is this more pronounced than in the field of forensic science. Top-quality scientific analysis is horrendously expensive, far beyond the reach of all but the deepest pockets, which means that most defendants, when confronted by the prosecution's vast army of government-funded expert witnesses, find themselves squashed into submission. The upshot of this imbalance is that, in the main, scientific testimony and conclusions go untested in court. Juries tend to pretty much accept what the experts say, and that's the end of it.

Because it is so costly, forensic science is generally reserved for the most serious cases, and this skews the scales even further, since most violent crimes are committed by the financially disadvantaged. When rich people find themselves in court, usually it's for some white-collar offense that requires little if any scientific testimony.

Which is what, in part, made the O. J. Simpson case so enthralling. Here was a murder defendant with seriously deep pockets, someone with the financial muscle to subject every syllable of forensic testimony to the most withering cross-examination. And let nobody doubt that he got full value for his money. But there was a kicker. Forensic science is supposed to take the guesswork out of crime detection. What took almost everyone by surprise here was the way in which statistics, graphs, state-of-the-art laboratories, percentages, probability theorems, and test-tube results could be so undermined by something the prosecution failed entirely to factor into the equation. We're talking about that great imponderable—human nature.

At about 10:20 P.M. on June 12, 1994, the sound of a dog barking disturbed the Sunday evening calm on South Bundy Drive, in the upscale Brentwood district of Los Angeles. It would be almost midnight before the whimpering Akita, its paws splashed with blood, led puzzled neighbors into number 875, where it lived. The reason for the animal's distress was instantly obvious. On the tiled outer pathway of the stucco condominium lay two bodies, almost floating in blood.

Police officers summoned to the scene found a woman sprawled face down. Her slashed throat had been opened up like a briefcase, nearly severing her head. To her right lay the body of a man. He, too, had been sliced to ribbons by a knife.

Within an hour it was established that the woman was probably Nicole Brown Simpson, the thirty-five-year-old owner of the condo and the ex-wife of O. J. Simpson, the former football great turned sportscaster/actor. The identity of the male victim remained a mystery.

As were the whereabouts of Orenthal James Simpson.

Over the next hour or so, numerous police officers signed the roster at the crime scene. The seventeenth officer to do so was someone who would play a critical role in this investigation, Detective Mark Fuhrman. Like all of his predecessors, Fuhrman did not disturb the bodies; he merely observed. He saw a number of objects adjacent to the dead man: a set of keys; a dark blue knit cap; a beeper; a blood-spattered white envelope; and a bloodstained left-hand leather glove lying just inches from Nicole's body. What looked like a bloody trail of footprints and spots led away from the bodies toward the back of the property.

When instructions were issued that Simpson should be told personally of the tragedy, Fuhrman volunteered that he knew where Simpson lived. As a patrol officer he had once visited Simpson's home, which was about two miles to the north.

Police visits to the Simpson residence on Rockingham Avenue were depressingly frequent. Simpson had an uncontrollable temper, and his flailing fists often left Nicole bruised and bloody. Oftentimes, fearful for her life, she had dialed 911 and begged the police to come and restrain her raging husband. By 1992 she'd suffered enough and moved out. That

October a divorce ended the seven-year marriage. The schism did nothing to curb Simpson's maniacal jealousy. The following year police officers were summoned to Nicole's new home after her ex-husband had shown up in a rage, kicked in the door, screamed obscenities, then started trashing her car. Over the years, official records listed no fewer than sixty-two separate incidents of physical and mental abuse by Simpson toward his wife.

And now she was dead.

A party of detectives set off in two separate cars for the five-minute drive north to the home of O. J. Simpson, which stood on the junction of Rockingham Avenue and Ashford Street. Outside the gated estate, facing north on Rockingham, was a 1994 white Ford Bronco, parked crazily, front wheels on the sidewalk, back end sticking out into the narrow street. After repeated attempts on the intercom buzzer failed to elicit a response, Fuhrman inspected the Bronco more closely, then summoned a fellow officer and pointed out what appeared to be blood near the driver's door handle.

Unable to raise the occupants by phone, the detectives decided to enter the property. With Nicole Simpson dead just two miles away and blood on the Bronco, they had reasonable cause to effect an entry without a warrant; this could well be another crime scene. Fuhrman clambered over the five-foot-high stone wall and unlocked the gate from the inside.

After getting no response at the front door, the detectives walked around the side of the house to a row of three guest bungalows. At the first one a man appeared. He identified himself as Brian "Kato" Kaelin, a friend and houseguest of Simpson. Arnelle Simpson, O. J.'s daughter from a previous marriage, was in the next bungalow, and she let the officers into the main house, which was deserted.

Kaelin told a strange story. The previous evening he and Simpson had returned to the house at 9:40 P.M. after eating out. Simpson had disappeared into the main house, and Kaelin decided to phone his girlfriend. At about 10:45 P.M., while still on the phone, Kaelin heard three loud banging noises coming from the rear of the building near the air-conditioning unit. Gut instinct told him there had been a minor earthquake, so he grabbed a flashlight and went outside to check for damage.

All he saw was a limousine parked outside the gate on Ashford. It had been ordered by Simpson to take him to Los Angeles International Airport to catch the red-eye to Chicago. A few minutes later O. J. appeared at the front door, and Kaelin helped Allan Park, the chauffeur, load cases into the trunk of the car, except for a small black bag that Simpson insisted on holding on to. Then the limousine left for LAX.

Detectives digested this information, then decided to call O. J. at his

Chicago hotel. When told the tragic news, his reaction was puzzling. Although seemingly distraught, he neglected to ask for any details regarding the death of his ex-wife. He said he would catch the first available flight back to L.A.

Moments later, Fuhrman, who had been alone, returned to the house after an absence of ten to fifteen minutes with news of a discovery in the garden. Behind Kaelin's bungalow he had found a bloodstained leather glove, which seemed to be a right-hand match to the one still lying in the garden back at South Bundy.

This was starting to look ominous, especially when blood drops were spotted in the driveway that led out the west gate onto Rockingham, then to the rear of the Bronco. Inside the vehicle were other red marks on the driver's door and on the console near the passenger's side. Another trail of blood spots led up to the front door of the house.

At 7:10 A.M. Dennis Fung, an LAPD criminalist, and his assistant, Andrea Mazzola, a trainee, arrived to begin the collection and documentation of evidence. Later that morning they would perform the same duties at Bundy, little dreaming of how those few hours would so profoundly affect their lives and this case.

News of the double murder had spread fast, and Bundy was now

O. J. Simpson's fall from grace.

swarming with reporters and cameras. To protect Nicole's body from prying telephoto lenses, a blanket was thrown over the corpse, an innocuous
courtesy that would have the most damaging repercussions.

Only now did the identity of the murdered man become known. His
driver's license identified him as twenty-five-year-old Ronald Goldman, a
waiter and friend of Nicole Simpson.

Later that day, when Simpson returned from Chicago, he was detained
and brought to Bundy in handcuffs. This was clearly improper, and he was
immediately released. As Tom Vannatter, the detective in charge of the
investigation, unlocked the cuffs, he noticed that the middle finger of
Simpson's left hand was bandaged. Simpson told a confused story of having first cut himself in Los Angeles, only to reopen the wound in Chicago,
then saying that he had sustained the injury on a broken glass in the bathroom sink of his Chicago hotel room. When detectives checked the room
at the O'Hare Plaza Hotel where Simpson had spent the previous night,
they did indeed find a broken glass in the bathroom sink.

Already, though, Simpson's verbal fuzziness had convinced everyone
that he had something to hide.

Bloodstained Socks in Bedroom

For comparison purposes, a sample of Simpson's blood was drawn and later
passed to Fung, who, at this time, was collecting evidence at Rockingham.
Among the items he and his partner bagged and tagged was a pair of navy
blue bloodstained socks in the master bedroom.

On June 14 Dr. Irwin Golden carried out autopsies on the two murder victims. Defensive wounds on Nicole's hands showed that she had
fought furiously for her life, until being almost decapitated by the slash-
ing wound across her neck. From the angle of the wounds, Golden determined that the attacker was probably right-handed and had slashed her
throat from behind, from left to right. Ronald Goldman appeared to have
been clubbed down from behind, then stabbed nineteen times in a frenzied attack.

Judging from the single set of bloody footprints leading away from the
bodies toward the back of the condominium, as well as similarities in the
pattern of wounds and mutilations on the two bodies, detectives were hunting a lone killer.

And that killer appeared to be O. J. Simpson. All week long the evidence against him multiplied exponentially:

1.   He had no alibi from 9:40 P.M. until 10:55 P.M., and smack in the middle of this time frame, Nicole's dog began barking, leading detectives to assume that this was when the murders were committed.

2.   There was a damaging statement from Allan Park, the chauffeur who had driven Simpson on the night of the murders. Instructed to arrive at the Rockingham estate no later than 10:45 P.M., he had arrived some twenty minutes early, but waited until 10:40 P.M. before calling the house on the Ashford gate intercom. There was no answer. Over the next ten minutes he tried repeatedly and without success to contact Simpson. Then, at approximately 10:50 P.M., he spotted a tall, well-built black man hurrying up the drive toward the house from the Rockingham gate side of the estate. Park now retried the buzzer. A man, who identified himself as Simpson, answered the phone, said he had been in the shower, and would be right down. Another ten minutes passed before Simpson appeared, clutching a bag and sweating profusely. Despite the cool night he insisted the air conditioning be kept on all the way to LAX.

3.   DNA typing of blood on the glove found at Rockingham confirmed that it was most likely a mixture of Simpson's and the two victims'.

4.   The gloves were identified as a pair belonging to O. J. Simpson.

5.   Blood found on the socks in the bedroom was typed as belonging to Nicole.

6.   Hairs found on Goldman's shirt and inside the knit cap discovered at the crime scene were found to be consistent with Simpson's hair, while hairs found on the Rockingham glove were compatible with Nicole's and Goldman's.

7.   Blue/black cotton fibers found on Goldman's shirt matched fibers in the socks found in Simpson's bedroom. Cashmere fibers removed from the knit hat matched fibers from the glove lining. Fibers in Simpson's Ford Bronco matched fibers found on the glove at Simpson's house and on the knit cap found at the crime scene.

8.   The distinctive waffle-type bloody footprints that led away from the crime scene at South Bundy were identified as being from a particular brand of Italian-made Bruno Magli shoes. Retailing for $160 per pair, they were sold by just forty stores across America. Only three hundred pairs of size 12 (Simpson's size) were ever sold. Initially Simpson had denied ever owing a pair. However, in September 1993, a press photographer had taken pictures of Simpson wearing these exact shoes at a stadium in New York.

Despite this avalanche of evidence, the DA's office trod carefully, wary of Simpson's vast lawyer purchasing power. They had to be certain. By June 19 they were convinced, and on that day a warrant for Simpson's arrest was issued. Except that the suspect was missing.

What happened next made for some of the most compelling images in TV history. There was something surreal about the spectacle of a slow-moving white Ford Bronco trundling sedately along an L.A. freeway, tracked by a posse of deferential patrol cars. It was hardly the stuff of a Hollywood action thriller. According to his friend at the wheel, Al Cowlings, Simpson had a .357 Magnum pressed to his own head, threatening suicide if stopped. He also had eight thousand dollars in cash, his passport, and a disguise kit consisting of a fake mustache and beard, and was presumably airport-bound, ready to skip the country—until being spotted.

After ninety minutes, the leisurely jaunt—it really couldn't be called a "chase"—ended when the Bronco rolled back into the driveway at Rockingham. While SWAT teams assumed their positions and ecstatic fans waved placards proclaiming "We love the Juice," O. J. holed up inside his mansion. Finally, after an hour of negotiation, the former football legend surrendered.

The next day he was charged with double murder.

Probably no murder defendant in California legal history had faced such an overwhelming forensic case. All twenty-five lawyers working the case for the district attorney's office were confident it was watertight. And in ordinary circumstances it would have been. But these were far from ordinary circumstances. Most defendants don't have ten million bucks in the bank. With that kind of war chest almost anything is possible.

Dubbed the "Dream Team" by a hyperbolic media, the eleven-man squad of attorneys assembled to keep Simpson out of prison was a curious mix. Initially, the point man had been Robert Shapiro—a renowned Hollywood fixer of celebrity "difficulties"—but as the trial date neared, this silky negotiator found himself shunted farther and farther into the background. Even the most seasoned murder trial attorney present, F. Lee Bailey, drifted off-camera as it became clear that Simpson's best—perhaps his only—chances of acquittal rested in the hands of two lawyers whose styles and personality could not have been more contradictory: Barry Scheck and Johnny Cochran.

Scheck is every inch the modern advocate. Well prepared and meticulous, Scheck, together with his partner Peter Neufeld, who also figured prominently in this trial, has built his reputation on the use and abuse of

DNA typing. He is no great courtroom orator in the tradition of Clarence Darrow or Earl Rogers; rather, his adversarial talents lie in the area of forensic nitpicking. He first numbs the senses with an avalanche of tedious arcana delivered in a drone that could anesthetize a roomful of insomniacs, then isolates a single inconsistency. Suddenly, he's up and running. Through endless repetition, circumlocutions, and shading, Scheck fights to transform this minor inconsistency from the tiniest molehill into an Everest of doubt. Naturally, this takes time. However, given Simpson's ability to keep the meter running, the feisty lawyer from Brooklyn had no clock ticking over his head. In short, Barry Scheck was free to do what he does best—bore and confuse juries into utter bewilderment.

With this object achieved, the defense baton was handed to the anchor leg, Johnny Cochran. Slick, polished, and sharper than a tack, this son of a onetime sharecropper knew instinctively how to convert Scheck's openings into real muscle. Cochran had fashioned a million-dollar career out of fighting racism in California law-enforcement agencies, and now with a jury made up of eight blacks, two of mixed descent, one Hispanic, and one white, the sly old campaigner wasn't about to switch ships. Let the scientists and experts bicker all they liked, Cochran knew deep in his gut how this particular battle was going to play out.

Well before the trial began, the defense leaked its strategy to a few favored journalists: O. J. had been the victim of a police frame-up—orchestrated by Detective Mark Fuhrman. After all, he'd found the blood on the Bronco; he'd found the bloodstained glove at Rockingham after an unregulated solo search; he'd previously tangled with Simpson and might possibly bear a grudge. Defense rumors circulated that Fuhrman had actually found the glove at Bundy, then transported it to Rockingham to incriminate his old adversary.

For this hypothesis to fly, the defense had to overcome the fact that Fuhrman was the seventeenth police officer to log in at the crime scene, almost two hours after the bodies were discovered, and that not one preceding officer had seen or reported more than the one glove found near the bodies.

Then there was the question of feasibility: Was Fuhrman really dumb enough to attempt to incriminate O. J. without first finding out if he had an alibi, or if some other suspect would enter the frame? And who in his or her right mind attempts to frame a multimillionaire international celebrity?

These were serious objections, or so one would imagine. But Cochran was sitting on an evidentiary zinger. He knew Fuhrman was a barefaced liar. And no jury likes that.

Simpson's Volcanic Temper

In her opening statement on January 24, 1995, Deputy District Attorney Marcia Clark portrayed Simpson as a man consumed by jealousy, a control freak who'd been rejected by the woman he'd loved, a homicidal volcano just waiting to blow. And on the night of June 13 Vesuvius erupted in Brentwood. In a carefully premeditated attack, Simpson took a knife—the murder weapon was never found—and butchered Nicole on her own doorstep. Poor Ronald Goldman, she explained, had merely been collateral damage, caught in the fallout of an ex-husband's lethal retribution.

Clark outlined the primary forensic evidence against the defendant, the blood spots that led from Nicole's condo to Simpson's home. "That trail of blood from Bundy through his own Ford Bronco and into his house in Rockingham is devastating proof of his guilt,"[1] she told the jury. Then there was the hair found on the knit cap that matched Simpson's hair; the socks in his bedroom that contained traces of the victims' blood; the cut on Simpson's hand. It all sounded so compelling.

The next day, Cochran tipped his hand early. Telling the jurors that the football legend was an innocent man, falsely accused by a prosecution out to win at any cost, he quoted from Martin Luther King Jr.: "Injustice anywhere is a threat to justice everywhere."[2] The evidence collected in this case was, he said, ". . . contaminated, compromised, and ultimately corrupted."[3] As for that "trail of blood," Cochran claimed that some of Simpson's blood sample had gone mysteriously missing before detectives passed the vial to the lab for processing, and it was this blood—allegedly 1.5 ml—that had been used to contaminate the socks in Simpson's bedroom.

At the conclusion of his address, Cochran made Simpson stand up and walk over to the jury box. Adding a decade with every step, Simpson hobbled on scarred and battered knees, prompting Cochran to query whether anyone in such pitiful shape was capable of killing anybody.

With the showboating over, it was time to get serious.

First came the police officers. Cochran tore into them, disputing not only their integrity but also their blinkered intransigence. He wanted to know if lead detective Tom Lange had ever considered "any other theory than that O. J. Simpson was the only perpetrator in this case."

"I had absolutely no other evidence that would point me in any other direction,"[4] Lange replied.

Whereupon Cochran threw in the name of Faye Resnick, a Beverly Hills socialite and friend of Nicole Simpson. Resnick's involvement with drugs was well known, and it allowed Cochran to advance the notion that Nicole and Goldman may have been killed by drug dealers bent on scaring Resnick into paying off her drug debts.

This was an absurdly weak premise, raising the specter of assassins who happened to shop at the same shoe and glove stores as O. J. Simpson and were of the same physical size as the defendant, but it did succeed in muddying the waters still further, reinforcing defense claims that the LAPD had pursued a single-minded vendetta against Simpson.

When Mark Fuhrman took the stand, Bailey bombarded him with allegations that he had planted evidence at Rockingham, charges that Fuhrman denied urbanely and repeatedly. Then came blunt accusations of pathological racism. Again Fuhrman smoothly denied the allegations.

"You say under oath that you have not addressed any black person as a nigger or spoken about black people as niggers in the past ten years, Detective Fuhrman?" Bailey asked.

"That's what I'm saying, sir,"[5] responded Fuhrman.

At that moment mental high-fives must have circled the crowded defense table. With that one sentence, they knew Fuhrman had handed O. J. the key to his cell door.

First, though, they needed to undermine the prosecution's expert witnesses.

Enter Barry Scheck.

His job was to spotlight what the criminalists hadn't done, rather than what they did. Piece by excruciating piece, he unraveled every aspect of the state's forensic case, puckering with displeasure over every isolated anomaly. When criminalist Dennis Fung took the stand, he could little have imagined that his ordeal would last three miserable weeks. An experienced professional, with eleven years on the job and more than five hundred crime scenes under his belt, nothing had prepared him for this kind of onslaught.

Scheck began with the blanket used to cover Nicole's body from prying cameras. What had been a simple humanitarian gesture to afford some kind of dignity in death now became a defense lifeline. He expressed anxiety about contamination, in particular the possibility that hairs on the blanket—hairs perhaps left by O. J. on a previous visit to the house—were transferred to the crime scene. Was it not, he asked, a "terrible mistake" to have used the blanket? Fung said it would depend on how clean it was.

"Well, if you had no idea how clean the blanket was, wouldn't it still be a terrible mistake to bring a blanket from inside the house right into the middle of a crime scene?"

"I would prefer that it was not done,"[6] Fung admitted.

Scheck kept wheedling away at minor operational blunders: a crime scene photograph that showed an ungloved hand holding the blood-spattered envelope that contained Nicole's eyeglasses; Fung's admission that he had only collected "representative samples"[7]of the blood in the Ford

Bronco, thus explaining why stains were still found in the vehicle six weeks after it was impounded; another confession from the witness that—contrary to textbook guidelines—he had placed blood samples into plastic bags as a temporary measure, even though this could foster bacteria growth and thereby distort any test results.

The exchanges really sizzled when Scheck accused Fung of lying to conceal when he actually received the vial of Simpson's blood. Battered and confused, the criminalist eventually recalled that he had given the vial of blood to his assistant, Andrea Mazzola. Scheck upped the ante, accusing Fung of substituting an original page in the crime scene checklist with a photocopy in order to change the stated time that he had received the blood vial. Even though this missing page subsequently turned up in a crime lab notebook, deflating defense claims of a conspiracy, by then the damage had been done.

On April 18 Dennis Fung finally limped from the stand. Such was his sense of relief that, to the astonishment of everyone present, he crossed over to the defense table and shook hands with Simpson and his lawyers. As a *coup de théâtre* it was magnificent. Moreover, it didn't cost Simpson a cent.

Earlier, Scheck had grilled Fung over why junior criminalist Andrea Mazzola had been allowed to collect most of the blood evidence. Now it was Mazzola's turn to come under the gun.

Her inquisitor was Peter Neufeld. Mazzola agreed that she had collected most of the blood samples without any supervision from Fung, although, in earlier hearings, Fung had claimed the opposite. Neufeld struggled to show that Mazzola did a sloppy job, using videotape of her resting a hand on a dirty footpath, wiping tweezers with a dirty hand, dropping several blood swabs. She admitted to mistakes in the evidence collection but denied that anyone had deliberately altered evidence. Like Fung she was hazy about the vial of Simpson's blood, fueling defense claims that the police had ample time to plant evidence.

Between them Scheck and Neufeld laid a smoke screen thick enough to obscure the damaging evidence, yet with enough gaps to focus on a few minor irregularities. It had been top-notch advocacy.

Greg Matheson, chief forensic chemist and the supervisor of the serology unit at the LAPD Special Investigations Department crime laboratory, testified that blood drops leading away from the murder scene were found to be consistent with Simpson's blood, pointing out that only one person in two hundred had that blood type. He also explained that the alleged shortfall on the vial of blood drawn from Simpson could have happened in a number of ways. Tests carried out indicated that tiny particles of blood could have stuck to gloves, laboratory equipment, or the pipettes—tube-like instruments used to withdraw blood from the vial—during the various forensic testing procedures.

One of the prosecution's big guns was expected to be Dr. Robin Cotton, laboratory director of Cellmark Diagnostics, one of the foremost DNA testing facilities in America. For three days Cotton explained the nature and function of DNA and the tests used to identify it. In determining the so-called genetic fingerprinting tests, two methods are used. One is restriction fragment length polymorphism (RFLP), and the other is polymerase chain reaction (PCR). The PCR test requires less blood as a sample, is quicker to perform, but is less reliable than the RFLP test.

There are three distinct steps in RFLP analysis: (1) processing of DNA from the suspect and the crime scene to produce autorads (X-ray films); (2) examination of the autorads to determine whether any sets of fragments match; and (3) if there is a match, determination of the match's statistical significance. Cotton claimed that DNA tests showed a genetic match between Simpson's blood and bloodstains leading away from the bodies at South Bundy. The odds that this stain could have come from anyone but Simpson, said Cotton, were about 1 in 170 million.

She also confirmed that blood on the socks found in the bedroom at Rockingham Avenue had the same genetic fingerprint as Nicole's, characteristics that matched only 1 in 9.7 billion, more than the population of the world.

Under cross-examination, Neufeld attacked Cellmark's statistical calculations and its DNA database. He pointed out that twice in the past, in 1988 and 1989, Cellmark had recorded false DNA matches. These came about through the eagerness of laboratories to stake a claim in the early days of the DNA gold rush. Sensing that huge profits were on tap from the new technology, some labs, including Cellmark, cut corners. But that was all in the past, Cotton insisted; in recent years Cellmark had passed more than one hundred quality-control tests.

Corroboration of Cotton's findings came from the California Department of Justice's DNA laboratory. Gary Sims testified that not only did the blood on the socks at Rockingham match Nicole's blood, but also that blood samples lifted from inside the Ford Bronco and from the glove found at Rockingham contained traces of Simpson, Nicole, and Ronald Goldman.

Taken in tandem, the statistical findings of Cellmark and the Department of Justice reached stratospheric levels, with the odds against the blood on the socks coming from anyone other than Nicole Simpson soaring to *1 in 21 billion!*

Scheck returned to the fray, engaging Sims in a wearisome dialogue about the accuracy or otherwise of DNA testing and statistical analysis. At times Scheck's questioning became so abstruse, his syntax so tangled, that trial judge Lance Ito asked Sims, "Do you understand the factors involved in the question?"

"About half of it,"[8] Sims sighed.

If the experts were struggling, what chance did the jury have? Sequestered already for 133 days, subjected to hour after hour of mind-numbing statistics and the finer distinctions of PCR and RFLP ad nauseam, their increasing bewilderment became all too obvious. In the end Judge Ito decided to revive sagging morale by authorizing a field trip to watch *Miss Saigon* and ride the Goodyear blimp.

The respite was welcome but brief. Then it was back on the same relentless treadmill. Prosecution experts came and went, bludgeoned by the dull hammer of Scheck's interrogation, and everyone struggled to stay awake.

The prosecution, too, was showing signs of fatigue. After their earlier mauling over procedural irregularities, they decided to switched tactics. Far better, they reasoned, to highlight discrepancies rather than have the defense do it for them. With this in mind when Los Angeles County Medical Examiner Dr. Lakshmanan Sathyavagiswaran testified, the prosecution wasted no time in conceding that Sathyavagiswaran's deputy, Irwin Golden, had made numerous errors during the autopsy. These included: Golden's failure to retain the contents of Nicole Simpson's stomach; his failure to record an injury on Nicole Simpson's brain; the mislabeling of a sample of Goldman's bile as urine; the failure to document several tears on Goldman's shirt and jeans; and the failure to take a palmprint from Nicole Simpson's left hand. None of these mistakes was critical, and none affected Sathyavagiswaran's opinion that Golden had reached fundamentally sound conclusions.

But an error is an error, and juries are notoriously susceptible to a carefully highlighted blunder. It creates doubt, and, like cancer, that doubt spreads, chewing up every bit of evidence it can find, until everything appears tarnished and worthless.

Which is what happened to the gloves.

Glove Evidence Backfires

They were supposed to be the jewel in the prosecution's crown. Both were stained with blood, with the Rockingham glove showing traces of Simpson and both victims. Manufactured by Aris Gloves from dark brown leather, they were cashmere-lined; size extra large; and, above all, exclusive.

Bloomingdale's, in New York City, was the only outlet, and between 1989 and 1992 the store sold just 240 pairs. On December 20, 1990, Nicole Simpson had purchased two pairs of these gloves. Press photographs and videotapes of Simpson showed him wearing this type of leather gloves during football telecasts in 1993 and 1994.

When assistant prosecutor Christopher Darden called Richard Rubin to the stand, the former vice president and general manager of Aris Gloves testified that he had measured Simpson's hand and estimated it to be size large to extra large. An earlier prosecution trial run with Detective Vannatter—whose hand size approximated that of Simpson's—revealed that identical gloves slipped on easily. Rather than wait for the defense to blindside him, Darden asked Simpson to try on the gloves.

Cochran now pulled off his masterstroke—he insisted that before attempting to try the brown leather gloves his client should don latex gloves to prevent any evidence transfer. Suitably shielded, Simpson then attempted to slip on the gloves.

Grimacing fit to bust, Simpson struggled vainly to get his hands inside the gloves, mouthing, "They're too tight."[9]

Darden must have wished that the ground would open up and swallow himself, the gloves, the trial, and the whole district attorney's office. This was a catastrophe.

Cochran did one hell of a job keeping the grin off his face. The trial still had three months to run, the prosecution had shot themselves in the foot with both barrels, and the defense still had its bombshell in the locker.

Through private investigators, the defense had gained access to some tape recordings made by a North Carolina writer researching racism in the LAPD. In these ten-year-old tapes Detective Mark Fuhrman could plainly be heard using the word "nigger." Having earlier denied under oath that he had ever used the insult, Fuhrman was now exposed as a perjurer of the grossest kind. Worse, the tapes were littered with gloating admissions that he and other officers had often planted evidence on suspects to secure convictions.

Although the predominantly black jury didn't get to hear all the tapes, they heard enough, and although not an iota of evidence existed to suggest that Fuhrman had acted improperly in this particular case, his credibility lay in tatters.

Mark Fuhrman sank beneath the waves and took the prosecution case with him.

Someone else whose reputation suffered a big hit in this case was defense witness Dr. Henry Lee, head of the Connecticut State Forensics Science Laboratory and one of the foremost forensic scientists in America. Enigmatic and studious, Lee delivers his testimony quietly and with admirable simplicity, the way a jury likes its science. This is not to say that he is always right. And in this case he appears to have made a real howler.

Thirteen days after the murder, Lee visited Bundy, where he found what looked like unidentified footprints on a piece of paper and on an envelope near the crime scene. Checking photographs of the scene, Lee testified that he also thought he detected what he called "partial parallel line imprint patterns,"[10] in Goldman's blood-soaked jeans and in the soil. From such a respected source this sounded like solid evidence to corroborate defense claims that more than one killer had been responsible for the attack.

All went swimmingly for Lee until William Bodziak, an FBI shoeprint expert who had testified earlier, was called back to the stand. Placing photos of the walkway taken on June 13 next to photos taken by Lee more than a week later, Bodziak demonstrated that a shoeprint and a second partial print identified by Lee were not visible in the earlier photos. It appeared as if the imprints were made after the crime scene was no longer secured by police, possibly by someone stepping in the water that was used to wash away the blood.

Bodziak's rebuttal clearly unnerved Lee. Back in his home state of Connecticut, he called a press conference at which he heaped effusive praise on the FBI forensic laboratory, then glumly declared that he had no intention of returning to the trial to defend his previous stance. "I feel disappointed by the whole process . . . my name will clear itself."[11]

This small victory aside, it had been a harrowing time for the prosecution. In her final arguments, Marcia Clark, still clinging to the moral high ground, begged the jury not to be deflected by Fuhrman's racism, to listen to their heads and heed the testimony of the forensics experts. "You have a wealth of evidence," she said. "And all of it is pointing to one person, the defendant."[12]

By contrast, Cochran took dead aim at the heart and gut. With the slickness of a street magician, he palmed O. J. Simpson, hid him from view, diverting everyone's attention instead to the verminous Fuhrman, who miraculously metamorphosed into an ideological bedfellow of Adolf Hitler. If Cochran's logic was shaky, then his passion was anything but. With the fervor of a fire-and-brimstone preacher he implored the jury to reject this "lying, genocidal, racist cop,"[13] and the police department that had shielded him. "You are empowered to say this is wrong. Stop this cover-up! Stop this cover-up!"[14]

It had been the longest trial ever held in California, costing more than

$200 million, running up 50,000 pages of trial transcript, with 150 witnesses. Forensic science testimony alone had occupied two months and included in excess of 10,000 references to DNA. Yet Cochran was able to demolish all of this with a single mantra centered on that infamous glove demonstration: "If it doesn't fit, you must acquit." Repeated endlessly during his final summation, these seven words drummed themselves into the jury's collective consciousness. Here was something tangible they could take with them to their deliberations.

Five hours were all they needed to decide that O. J. Simpson was not guilty.

It is hard to recall any other trial that has engendered such rancor and acrimony. The backbiting was spectacular. Prosecutors had gambled the farm on forensic science, only to learn the painful lesson that it can be a double-edged sword. The more science one introduces, the more compelling it can appear to a jury, but, by the same token, every addition opens up one more potential defense loophole. If every defendant had O. J.'s financial resources and, above all, time, the conviction rate would plummet close to zero; courts would be throttled, and juries would be stupefied into delivering "not guilty" verdicts.

A seemingly unassailable forensic case had been destroyed by the raw power of human emotion. After the verdict was safely in, even fellow defense lawyers tried to distance themselves from Cochran's crude exploitation of the jury's racial makeup. But it's difficult to see what other ploy he could have adopted. His duty was to his client, not the mores of mainstream America. Besides, he wasn't doing anything new.

For decades, good ol' boy lawyers had been pulling off the same stunt in courtrooms all across Dixie, winning unfathomable verdicts from redneck juries for defendants accused of murdering African Americans. All Cochran did was switch the scenario. Suddenly this jury had a one-in-a-lifetime chance to tip the scales the other way, an opportunity to exercise the ultimate civil right—acquit a man whom most of the world still believes to be a calculating and callous double murderer.*

---

*At a subsequent civil trial, Simpson was adjudged to have been responsible for the deaths of Nicole Simpson and Ronald Goldman and ordered to pay $33.5 million in punitive and compensatory damages.

Notes

Chapter 1. The Turin Shroud (1355)

1. *Albuquerque Journal* (October 22, 1988).
2. *The D'Arcis Memorandum*, p. 11, 1389.
3. Ibid.
4. Ian Wilson, *The Evidence of the Shroud* (London: O'Mara, 1986), p.12.
5. Ibid.
6. *Biblical Archaeology Review*, 24, no. 26 (1998).
7. Ibid.
8. Ibid.
9. Tom Tullett, *Clues to Murder* (London: Bradley Head, 1986), p. 220.
10. Ibid., p. 222.
11. Ibid.
12. Ibid.
13. Ibid.
14. *Time* (April 20, 1998).
15. Ibid.
16. Ibid.
17. Robert Hedges, "A Note Concerning the Application of Radiocarbon Dating to the Turin Shroud," *Approfordimento Sindone*, 2 (1998): 1.
18. Ibid., 1998.
19. *The Mission* (Spring 1996).
20. *Sunday Telegraph* (London) (November 15, 1998).
21. *Daily Telegraph* (London) (June 18, 1999).
22. Wilson, *The Evidence of the Shroud*, p. 29
23. Ibid.

Chapter 2. Napoleon Bonaparte (1821)

1. *Crimes and Punishment*, vol. 14. (London: BPC Publishing, 1974), p. 102.
2. Thomas I. Noguchi and Joseph DiMona, *Coroner at Large* (New York: Pocket Books, 1986), pp. 203–204.
3. Weider lecture (February 18, 1998).
4. Ibid.
5. Ibid.
6. David Hooper, *Vaishnava News* (September 16, 1999).

7. Bob Elmer, *Journal of the Association of Friends of the Waterloo Committee*. www.afwc.ic24.net.
8. Philip Corso and Thomas Hindmarsh, *Journal of the History of Medicine* (London: Oxford University Press, 1996), pp. 89–96.
9. Ibid.
10. J. Oh Shin, *Electrophysiological Profile in Arsenic Neuropathy*. 54 (1991): 1103–1105.
11. *Ottawa Citizen* (September 1, 1998).
12. Michael Baden and Judith Adler Hennessee, *Unnatural Death* (London: Sphere, 1991), p. 40.
13. *London Daily Telegraph* (June 2, 2001).
14. *Ottawa Citizen* (September 1, 1998).
15. Ibid.

Chapter 3. Alfred Packer (1874)

1. *Times* (London) *Magazine* (June 16, 2001).
2. *Saguache Chronicle* (March 23, 1883).
3. Packer's second confession, Colorado State Archive (March 16, 1883).
4. Ibid.
5. *Denver Post* (July 16, 1989).
6. Colorado State Archive.
7. *Minnesota Daily Online* (October 26, 1994).
8. *Real Life Crime*, vol. 14 (London: Eaglemoss, 1993), p. 304.
9. Ibid.
10. Ibid., p. 305.
11. Ibid.
12. Ibid.
13. Ibid., p. 307.
14. Ibid.
15. *Denver Post* (April 26, 1907).
16. Ibid. (February 5, 2001).
17. National Public Radio, *Weekend Edition* (February 17, 2001).
18. *Denver Post* (February 13, 2001).
19. National Public Radio, *Weekend Edition* (February 17, 2001).
20. *Denver Post* (February 13, 2001).

Chapter 4. Donald Merrett (1926)

1. Jurgen Thorwald, *Dead Man Tell Tales* (London: Pan Books, 1968) p. 114.
2. Ibid., p 115.

3. Ibid.
4. *RLC*, vol. 22 (1999), p. 486.
5. Ibid.
6. Ibid.
7. Jack House, *Murder Not Proven* (Glasgow: Richard Drew Publishing, 1984), p. 147.
8. Thorwald, p. 119.
9. Ibid., p. 120.
10. Sir Sydney Smith, *Mostly Murder* (London: Grafton, 1984), p. 177.
11. Ibid., p 178.
12. Thorwald, p. 125.
13. Ibid., p. 128.
14. *RLC*, p. 491.
15. Douglas G. Browne and E. V. Tullet, *Bernard Spilsbury* (London: George G. Harrap, 1941), p. 363.
16. Ibid.
17. Thorwald, p. 129.
18. Ibid.
19. Ibid.
20. Browne and Tullet, p. 364.
21. Ibid., p. 365.
22. Smith, p. 180.
23. Ibid., p. 182.
24. Keith Simpson, *Forty Years of Murder* (London: George G. Harrap, 1978), p. 77.
25. Ibid., p. 79.

Chapter 5. William Lancaster (1932)

1. *Times* (London) (July 11, 2001).
2. Ibid.
3. *Miami Herald* (October 1, 1989).
4. Ibid.
5. Ibid.
6. Ibid.
7. Ibid.
8. Ralph Barker, *Verdict on a Lost Flyer* (London: George G. Harrap, 1969), p. 101.
9. Ibid., p. 102.
10. Ibid., p. 107.
11. *Miami Herald* (October 1, 1989).

12. Ibid.
13. Ibid.
14. Barker, p. 109.
15. Ibid., p. 119.
16. Ibid., p. 179.
17. Ibid., p. 176.
18. Ibid., p. 179.
19. Ibid., p. 180.
20. Ibid.
21. Ibid.
22. Ibid., p. 114.
23. *Fort Lauderdale Sun-Sentinel* (August 19, 1932).
24. Barker, p. 234.
25. Ibid., p. 201.
26. Ibid., p. 187.

Chapter 6. Sir Henry John Delves Broughton (1941)

1. *RLC*, vol. 29, p. 628.
2. Ibid.
3. Ibid.
4. Ibid.
5. Ibid.
6. Ibid., p. 629.
7. James Fox, *White Mischief* (London: Penguin, 1984), p. 82.
8. Julian Symons, *A Reasonable Doubt* (London: Cresset Press, 1960), p. 199.
9. Fox, p. 87.
10. Ibid., p. 88.
11. *RLC*, p.627.
12. Fox, p. 95.
13. *RLC*, p. 631.
14. Ibid., p. 632.
15. Symons, p. 206.
16. Ibid.
17. Ibid., p. 207.
18. Ibid.
19. Fox, p. 111.
20. Symons, p. 208.
21. *Sunday Times* (August 9, 1998).
22. Ibid.

Chapter 7. Alfred de Marigny (1943)

1. *New York Times* (October 20, 1941).
2. Peter Fearon, *Behind the Palace Walls* (Secaucus, N.J.: Carol Publishing, 1996), p. 157.
3. Alfred De Marigny and Herskovitz, *A Conspiracy of Crowns* (New York: Bantam, 1988), p. 19.
4. James Leasor, *Who Killed Sir Harry Oakes?* (London: Heinemann, 1983), p. 4.
5. Ibid., p. 45.
6. Ibid., p. 46.
7. Ibid.
8. De Marigny, p. 50.
9. Ibid., p. 51.
10. Ibid., p. 52.
11. Leasor, p. 29.
12. Ibid., p. 50.
13. Ibid.
14. Ibid.
15. Ibid., p. 54.
16. Ibid., p. 58–59.
17. Ibid., p. 60.
18. Ibid.
19. Ibid., p. 73.
20. Ibid., p. 54.
21. De Marigny, p. 263.

Chapter 8. Samuel Sheppard (1954)

1. *Cleveland Plain Dealer* (March 7, 2000).
2. F. Lee Bailey and Harvey Aronson, *The Defense Never Rests* (New York: Signet, 1971), p. 70.
3. Ibid., p. 71.
4. Dorothy Kilgallen, *Murder One* (New York: Random House, 1967), p. 246.
5. Ibid., p. 302.
6. Ibid., p. 262.
7. Ibid., p. 260.
8. Bailey and Aronson, p. 93.
9. Ibid., p. 101.
10. Ibid.

11. Ibid., p. 109.
12. *Cleveland Plain Dealer* (April 5, 2000).
13. Ibid. (February 26, 2000).
14. Ibid. (February 26, 2000).
15. *Court TV* (February 14, 2000).
16. Ibid. (February 29, 2000).
17. Ibid.
18. Ibid.
19. Ibid. (October 6, 1999).
20. Ibid. (February 24, 2000).
21. Ibid. (February 14, 2000).
22. Ibid.

Chapter 9. Steven Truscott (1959)

1. Robert Jackson, Steven Truscott (1959), *Francis Camps* (London: Hart-Davis MacGibbon, 1975), p. 28.
2. Brian Lane, *Encyclopedia of Forensic Science* (London: Headline, 1992), p. 619.
3. Simpson, p. 282.
4. Baden and Hennessee, p. 103.
5. *Autopsy Report* (June 11, 1959).
6. Simpson, p. 283.
7. Ibid., p. 280.
8. Isabel LeBourdais, *The Trial of Steven Truscott* (Toronto: McClelland & Stewart, 1996), p. 44.
9. Ibid., p. 50.
10. Simpson, p. 279.
11. Ibid., p. 284.
12. Ibid., p. 285.
13. LeBourdais, p. 28.
14. *Toronto Daily Star* (October 6, 1966).
15. Ibid. (October 13, 1966).
16. Simpson, p. 287.
17. Ibid., p. 286.
18. Ibid., p. 288.
19. Ibid.
20. *Toronto Daily Star* (October 6, 1966).
21. Ibid.
22. *the fifth estate* (March 29, 2000).
23. Ibid.

24. Helpern, p. 151.
25. *the fifth estate* (March 29, 2000).
26. Simpson, p. 45.

Chapter 10. Lee Harvey Oswald (1963)

1. CBS (May 6–7, 1998).
2. *Times* (March 16, 2001).
3. *Warren Commission Report* (Washington, D.C.: U.S. Government Printing Office, 1964), p. 182.
4. Ibid., p. 592.
5. Vincent J. M. Di Maio, *Gunshot Wounds: Practical Aspects of Firearms, Ballistics, and Forensic Techniques* (New York: Elsevier Science Publishing, 1985), p. 267.
6. *Science* (October 1982).
7. Ibid.
8. *Warren Commission Report*, p. 77.
9. Ibid.
10. Ibid., p. 86.
11. Ibid., p. 87.
12. Ibid., pp.183–187.
13. Ibid., p. 105.
14. Henry Hurt, *Reasonable Doubt* (New York: Holt, Rinehart, & Winston, 1985), p. 99.
15. *Warren Commission Report*, p. 97.

Chapter 11. Jeffrey MacDonald (1970)

1. *Fayetteville Observer* (February 17, 2000).
2. Joe McGinnis, *Fatal Vision* (New York: Signet, 1984), p. 15.
3. Ibid., p. 21.
4. Fort Bragg army report (February 20, 1970).
5. McGinnis, p. 197.
6. "Family Killers," *Court TV* (1999).
7. *Dick Cavett Show* (December 15, 1970).
8. McGinnis, p. 464.
9. Ibid., p. 498.
10. Ibid., pp. 149–150.
11. Ibid., p. 151.
12. Ibid.
13. Ibid., p. 511.

14. "Family Killers," *Court TV* (1999).
15. Ibid.
16. Noguchi, p. 83.
17. Ibid.

Chapter 12. Lindy Chamberlain (1980)

1. *Times* (London) (December 5, 2000).
2. John Bryson, *Evil Angels* (New York: Viking Penguin, 1985), p. 40.
3. Ibid., p. 244.
4. Ibid.
5. Tullett, p. 13.
6. Bryson, p. 308.
7. Ibid., p. 310.
8. Ibid., p. 317.
9. Ibid., p. 317.
10. Ibid., p. 428.
11. Ibid., p. 432.
12. Edward W. Knappman, ed., *Great World Trials* (Detroit: Visible Ink, 1997), p. 409.
13. Ibid.
14. Ibid.
15. Ibid., p. 410.

Chapter 13. Roberto Calvi (1982)

1. Rupert Cornwell, *God's Banker* (London: Gollancz, 1983), p. 169.
2. Ibid., p. 196.
3. Noguchi, p. 255.
4. Simpson, p. 318.
5. Ibid., p. 318.
6. Noguchi, p. 268.
7. Jasper Ridley, *The Freemasons* (London: Constable, 1999), p. 273.
8. Noguchi, p. 259.
9. Ibid.
10. Ibid., p. 258.
11. Cornwell, p. 191.
12. Noguchi, p. 256.
13. *Daily Telegraph* (London) (July 24, 1982).
14. Ibid.
15. Noguchi, p. 261.

16. Ibid.
17. Ibid.
18. Ibid.
19. Cornwell, p. 204.
20. *Daily Telegraph* (London) (June 15, 1983).
21. Ibid. (June 23, 1983).
22. Ibid. (June 24, 1983).
23. Cornwell, pp. 248–249.
24. *Daily Telegraph* (London) (June 28, 1983).
25. Noguchi, pp. 263–264.
26. *Times* (London) (December 10, 2000).
27. Ibid.
28. Ibid.
29. *La Stampa* (June 15, 1982).

Chapter 14. Colin Stagg (1992)

1. "Rachel Nickell Story," television broadcast (June 27, 2001).
2. Paul Britton, *The Jigsaw Man* (London: Corgi, 1998), p. 247.
3. Ibid., p. 250.
4. "Rachel Nickell, the untold story, 1994," television broadcast (September 20, 1994).
5. Ibid.
6. Britton, p. 260.
7. Ibid.
8. Ibid., p. 330.
9. Ibid., p. 338.
10. Ibid., p. 349.
11. Ibid., p. 362.
12. "World in Action," television broadcast (September 19, 1994).
13. Ibid.
14. Ibid.
15. *The Observer* (February 21, 1999).
16. Ibid.
17. "World in Action."
18. Ibid.
19. *Guardian* (Manchester) (September 15, 1994).
20. Britton, p. 536.
21. *Independent* (September 15, 1994).
22. "World in Action."
23. Ibid.

24. *Daily Telegraph* (London) (May 5, 1995).
25. *Guardian* (Manchester) (September 2, 1999).
26. *Times* (London) (November 12, 2000).
27. "World in Action."

Chapter 15. O. J. Simpson (1994)

1. Trial testimony (TT) (January 24, 1995).
2. *Atlantic Monthly* (August 1963).
3. TT (January 30, 1995).
4. Ibid. (March 6, 1995).
5. Ibid. (March 15, 1995).
6. Ibid. (April 5, 1995).
7. Ibid. (April 11, 1995).
8. Ibid. (May 22, 1995).
9. Ibid. (June 15, 1995).
10. Ibid. (August 23, 1995).
11. *San Francisco Chronicle* (September 16, 1995).
12. TT (September 29, 1995).
13. *San Francisco Chronicle* (September 28, 1995).
14. TT (September 27, 1995).

Bibliography

Chapter 1. The Turin Shroud (1355)

Garza-Valdes, Leoncio A. *The DNA of God?* New York: Doubleday, 1999.

Heller, John H. *Report on the Shroud of Turin.* Boston: Houghton Mifflin, 1983.

Nickell, Joe. *Inquest on the Shroud of Turin.* Buffalo, N.Y.: Prometheus Books, 1983.

Sox, H. David. *The Image on the Shroud.* London: Unwin, 1981.

Stevenson, Kenneth E., and Gary R. Habermas. *Verdict on the Shroud.* London: Hale, 1982.

Tullett, Tom. *Clues to Murder.* London: Bodley Head, 1986.

Wilson, Ian. *The Turin Shroud.* London: Penguin, 1979

————. *The Evidence of the Shroud.* London: O'Mara, 1986.

————. *The Blood and the Shroud.* New York: Free Press, 1998.

Zugibe, Frederick T. *The Cross and the Shroud.* Garnerville, N.Y.: Angelus Books, 1982.

Chapter 2. Napoleon Bonaparte (1821)

Baden, Michael, and Judith Adler Hennessee. *Unnatural Death.* London: Sphere, 1991.

Crimes and Punishment. Vol. 14. London: BPC Publishing, 1974.

Encyclopaedia Britannica, 2001 (C.D.)

Forshufvud, Sten. *Who Killed Napoleon?* London: Hutchinson, 1962.

Forshufvud, Sten, and Ben Weider. *Assassination at St. Helena.* Vancouver, B.C., Canada: Mitchell Press, 1978.

Noguchi, Thomas T., and Joseph DiMona. *Coroner at Large.* New York: Pocket Books, 1986.

Smyth, Frank. *Cause of Death.* London: Pan Books, 1982.

Weider, Ben, and Sten Forshufvud. *Assassination at St. Helena Revisited.* New York: Wiley, 1995.

Chapter 3. Alfred Packer (1874)

Crimes and Punishment. Vol. 17. London: BPC Publishing, 1974.

Gant, Paul H. *The Case of Alfred Packer.* Denver: University of Denver Press, 1957.

Kushner, Ervan F. *Alferd G. Packer, Cannibal! Victim?* Frederick, Colo.: Platte 'n Press, 1980.

Real Life Crimes. Vol. 14. London: Eaglemoss Publications, 1993.

Chapter 4. Donald Merrett (1926)

Browne, Douglas G., and E. V. Tullett. *Bernard Spilsbury.* London: George G. Harrap, 1941.

Crimes and Punishment. Vol. 6. London: BPC Publishing, 1974.

Cuthbert, C. R. M. *Science and the Detection of Crime.* London: Hutchinson, 1958.

Gaute, J. H. H., and Robin Odell. *The New Murderers' Who's Who.* New York: Dorset Press, 1979.

House, Jack. *Murder Not Proven.* Glasgow: Richard Drew Publishing, 1984.

Lane, Brian. *Encyclopedia of Forensic Science.* London, Headline, 1992.

Real Life Crimes. Vol. 22. London: Eaglemoss, 1993.

Skelton, Douglas. *Blood on the Thistle.* Edinburgh: Mainstream Publishing, 1992.

Smith, Sir Sydney. *Mostly Murder.* London: Grafton, 1984.

Symons, Julian. *A Reasonable Doubt.* London: Cresset Press, 1960.

Thorwald, Jürgen. *Dead Men Tell Tales.* London: Pan Books, 1968.

Wilson, Colin, and Patricia Pitman. *Encyclopedia of Murder.* New York: G. P. Putnam's Sons, 1962.

Chapter 5. William Lancaster (1932)

Barker, Ralph. *Verdict on a Lost Flyer.* London: George G. Harrap, 1969.

Crimes and Punishment. Vol. 17. London: BPC Publishing, 1974.

Gaute, J. H. H., and Robin Odell. *The New Murderers' Who's Who.* New York: Dorset Press, 1979.

Chapter 6. Sir Henry John Delves Broughton (1941)

Bennett, Benjamin. *Genius for the Defense.* Cape Town: Harold Timmins, 1959.

————. *Who Shot the Earl of Erroll?* London: Guild Publishing, 1990.

Crimes and Punishment. Vol. 5. London: BPC Publishing, 1974.

Fox, James. *White Mischief.* London: Penguin, 1984.

Gaute, J. H. H., and Robin Odell. *The New Murderers' Who's Who.* New York: Dorset Press, 1979.

————. *Murder "Whatdunit."* London: George G. Harrap, 1982.

Real Life Crimes. Vol. 29. London: Eaglemoss, 1993.

Symons, Julian. *A Reasonable Doubt.* London: Cresset Press, 1960.

Tyzebinski, Errol. *Dead Reckoning.* London: Fourth Estate, 2000.

Chapter 7. Alfred de Marigny (1943)

Crimes and Punishment. Vol. 5. London: BPC Publishing, 1974.

DeMarigny, Alfred, and Mickey Herskovitz. *A Conspiracy of Crowns.* New York: Bantam, 1988.

Fearon, Peter. *Behind the Palace Walls.* Secaucus, N.J.: Carol Publishing, 1996.

Gaute, J. H. H., and Robin Odell. *The New Murderers' Who's Who.* New York: Dorset Press, 1979.

————. *Murder "Whatdunit."* London: George G. Harrap, 1982.

Leasor, James. *Who Killed Sir Harry Oakes?* London: Heinemann, 1983.

Odell, Robin. *Science against Crime.* London: Marshall Cavendish, 1982.

Smyth, Frank. *Cause of Death.* London: Pan Books, 1982.

Symons, Julian. *A Reasonable Doubt.* London: Cresset Press, 1960.

Chapter 8. Samuel Sheppard (1954)

Bailey, F. Lee, and Harvey Aronson. *The Defense Never Rests.* New York: Signet, 1971.

Crimes and Punishment. Vol. 3. London: BPC Publishing, 1974.

Gaute, J. H. H., and Robin Odell. *The New Murderers' Who's Who.* New York: Dorset Press, 1979.

————. *Murder "Whatdunit."* London: George G. Harrap, 1982.

Holmes, Paul. *The Sheppard Murder Case.* New York: David McKay, 1961.

Kilgallen, Dorothy. *Murder One.* New York: Random House, 1967.

Knappman, Edward W., ed. *Great American Trials.* Detroit: Visible Ink, 1994.

Pollack, Jack Harrison. *Dr. Sam—An American Tragedy.* Chicago: Regnery, 1972.

Sheppard, Sam. *Endure and Conquer.* Cleveland: World, 1966.

Smyth, Frank. *Cause of Death.* London: Pan Books, 1982.

Chapter 9. Steven Truscott (1959)

Baden, Michael, and Judith Adler Hennessee. *Unnatural Death.* London: Sphere, 1991.

Crimes and Punishment. Vol. 13. London: BPC Publishing, 1974.

Helpern, Milton, and Bernard Knight. *Autopsy.* London: George G. Harrap, 1979.

Lane, Brian. *Encyclopedia of Forensic Science.* London: Headline, 1992.

LeBourdais, Isabel. *The Trial of Steven Truscott.* Toronto: McClelland & Stewart, 1966.

Simpson, Keith. *Forty Years of Murder.* London: George G. Harrap, 1978.

Trent, Bill, and Steven Truscott. *Who Killed Lynne Harper?* Vancouver, B.C., Canada: Optimum, 1979.

Truscott, Mary R. *Brats.* New York: E. P. Dutton, 1989.

Chapter 10. Lee Harvey Oswald (1963)

Di Maio, Vincent J. M. *Gunshot Wounds: Practical Aspects of Firearms, Ballistics, and Forensic Techniques.* New York: Elsevier Science Publishing, 1985.

Garrison, Jim. *On the Trail of the Assassins.* New York: Warner, 1988.

Hurt, Henry. *Reasonable Doubt.* New York: Holt, Rinehart, & Winston, 1985.

Marrs, Jim. *Crossfire: The Plot that Killed Kennedy.* New York: Pocket Books, 1989.

Meagher, Sylvia. *Accessories After the Fact.* New York: Vintage Books, 1967.

Posner, Gerald. *Case Closed?* New York: Random House, 1993.

Summers, Anthony. *Conspiracy.* London: Gollancz, 1980.

Warren Commission Report. Washington, D.C.: U.S. Government Printing Office, 1964.

Wecht, Cyril, and Mark Curriden, Mark, Benjamin Wecht. *Cause of Death.* New York: E. P. Dutton, 1993.

Chapter 11. Jeffrey MacDonald (1970)

Bost, Fred, and Jerry Potter. *Fatal Justice.* New York: W. W. Norton, 1995.

Garbus, Martin. "McGinnis: A Travesty of Libel." *Publishers Weekly* (April 21, 1989).

Gaute, J. H. H., and Robin Odell. *The New Murderers' Who's Who.* New York: Dorset Press, 1979.

Knappman, Edward W., ed. *Great American Trials.* Detroit: Visible Ink, 1994.

Malcolm, Janet. *The Journalist and the Murderer.* New York: Alfred A. Knopf, 1990.

McGinnis, Joe. *Fatal Vision.* New York: Signet, 1984.

Noguchi, Thomas T., and Joseph DiMona. *Coroner at Large.* New York: Pocket Books, 1986.

Taylor, John. "Holier than Thou." *New York Times* (March 27, 1989).

Chapter 12. Lindy Chamberlain (1980)

Bryson, John. *Evil Angels.* New York: Viking Penguin, 1985.

Chamberlain, Lindy. *Through My Eyes.* London: Heinemann, 1991.

Gaute, J. H. H., and Robin Odell. *The New Murderers' Who's Who.* New York: Dorset Press, 1979.

Knappman, Edward W., ed. *Great World Trials.* Detroit: Visible Ink, 1997.

Tullett, Tom. *Clues to Murder.* London: Bodley Head, 1986.

Wilson, Colin, and Donald Seaman. *The Encyclopedia of Modern Murder.* New York: G. P. Putnam's Sons, 1983.

Chapter 13. Roberto Calvi (1982)

Cornwell, Rupert. *God's Banker*. London: Gollancz, 1983.

DiFonzo, Luigi. *St. Peter's Banker*. New York: Watts, 1983.

Gurwin, Larry. *The Calvi Affair*. London: Macmillan, 1983.

Noguchi, Thomas T., and Joseph DiMona. *Coroner at Large*. New York: Pocket Books, 1986.

Raw, Charles. *The Moneychangers*. London: Harvill, 1992.

Ridley, Jasper. *The Freemasons*. London: Constable, 1999.

Simpson, Keith. *Forty Years of Murder*. London: George G. Harrap, 1978.

Tosches, Nick. *Power on Earth*. New York: Arbor House, 1986.

Chapter 14. Colin Stagg (1992)

Britton, Paul. *The Jigsaw Man*. London: Corgi, 1998.

Fielder, Michael. *Killer on the Loose*. London: Blake, 1994.

Hanscombe, André. *The Last Thursday in July*. New York: Century, 1996.

Chapter 15. O. J. Simpson (1994)

Bosco, Joseph. *A Problem of Evidence*. New York: William Morrow, 1996.

Bugliosi, Vincent. *Outrage*. New York: W. W. Norton, 1996.

Dershowitz, Alan M., and Alan Morton. *Reasonable Doubts*. New York: Simon & Schuster, 1996.

Fuhrman, Mark. *Murder in Brentwood*. Washington, D.C.: Regnery, 1997.

Schmalleger, Frank. *Trial of the Century*. Upper Saddle River, N.J.: Prentice Hall, 1996.

Index